Cocktails, Mocktails and Garnishes from the Garden

Cocktails, Mocktails and Garnishes from the Garden

Recipes for Beautiful Beverages with a Botanical Twist

Katie Stryjewski

yellow pear press

An imprint of Mango Publishing

CORAL GABLES

Cover Design: Elina Diaz
Cover Photo/illustration: Katie Stryjewski
Layout & Design: Elina Diaz

For permission requests, please contact the publisher at:
Mango Publishing Group
2850 S Douglas Road, 2nd Floor
Coral Gables, FL 33134 USA
info@mango.bz

For special orders, quantity sales, course adoptions and corporate sales, please email the publisher at sales@mango.bz. For trade and wholesale sales, please contact Ingram Publisher Services at customer.service@ingramcontent.com or +1.800.509.4887.

Cocktails, Mocktails, and Garnishes from the Garden: Recipes for Beautiful Beverages with a Botanical Twist

Library of Congress Cataloging-in-Publication number: 2020950843
ISBN: (print) 978-1-64250-496-5 , (ebook) 978-1-64250-497-2
BISAC category code: CKB006000, COOKING / Beverages / Alcoholic / Bartending & Cocktails

Printed in the United States of America

Table of Contents

Rum 133

Tequila 151

Introduction

For me, cocktails have always been a little bit magical. For years, they seemed like mysterious concoctions of strange and exotic ingredients, and bartenders were mystical alchemists who knew the secrets of their preparation. Even as I learned how to make them myself and that mystique faded, they never lost their magic. A cocktail makes special occasions more special, makes guests feel welcome, and stimulates the best conversation between friends. Classic recipes and historic spirits connect us to the past in a concrete and tangible way. Cocktails are special.

I actually remember the exact evening, in fact the exact *drink*, that made me fall in love with craft cocktails. It was 2009, and I found out that I had been awarded a fellowship for graduate school. My husband and I planned a big night out to celebrate. We went to a fancy cocktail bar for drinks, something we couldn't afford to do regularly. Based on the menu's description, I chose an Aviation, which I now know is a classic cocktail made with gin, lemon juice, maraschino liqueur, and crème de violette. It arrived in an appropriately fancy glass, a lavender drink with a brandied cherry sitting at its center. I'd never had crème de violette or maraschino liqueur before. They were like nothing I'd ever tasted. All the components of the drink came together into a perfect, harmonic whole. I was converted.

Because we couldn't afford to go out for drinks often, I decided to try to learn to make good cocktails at home. The Aviation led me toward other classics like the Old Fashioned,

Tom Collins, and Pegu Club, as well as to blogs that posted new recipes from local bars. Thanks to those bloggers, I could go out and have a drink I loved, and then find out what was in it and add those things to my bar. My collection of bottles and recipes slowly grew.

I started my own blog, Garnish, in 2015. Since I was trying to learn about spirits and cocktails from scratch, and since I was finding other blogs so helpful in the process, I thought it could be interesting to document the process and try to put the information out there for other people to find. I began working my way through different ingredients and classic cocktails, posting recipes and photos and researching the history of each drink. I found it fascinating—often more fascinating than my actual graduate school research! Nothing makes a cocktail more enjoyable than understanding how it connects you to the past.

To really understand the concept of "craft cocktails," you have to start in the 1800s. Today's trends have their roots in this period, and there's an effort to emulate and venerate it in many ways—as is evidenced by the current stereotype of the suspender-clad, bearded bartender.

Prior to the nineteenth century, it wasn't the fashion to order individual cocktails. Instead, drinkers would order a bowl of punch to share at their own table that a bartender would mix up behind the scenes. This changed in the early days of the United States. In colonial times, people didn't just drink at taverns in the evening—they consumed alcohol with sugar and bitters at all times of the day, often for its purported health benefits. These individual health tonics evolved into the cocktails we know and love today. And with them, an entire culture of bars and bartending arose, a distinctly American innovation. Much of what we know about this period comes from a famous and flamboyant bartender named Jerry Thomas, who wrote a bartender's guide called *How to Mix Drinks or The Bon Vivant's Companion* in 1862. It contains the first known printed recipes for many drinks.

Jerry Thomas ushered in a golden era of cocktails, when many recipes that are now classics were first created and served. This lasted until Prohibition, which drove America's cocktail culture underground. Though this is now romanticized by modern speakeasies and a love for Prohibition-era style, the craft cocktail never really recovered. The rest of the twentieth century was a time of vodka, pre-packaged mixers, and chain restaurant bars. It's an era that is often referred to as the cocktail's dark ages.

But like the medieval dark ages, this one was followed by a renaissance. In the '80s and '90s, big names like Dale DeGroff, David Wondrich, and Jeff Berry began reading about, writing about, and serving classic cocktails. By the early 2000s, iconic bars like Milk & Honey, Flatiron Lounge, and Pegu Club were hand-carving ice and filling coupe glasses with pre-Prohibition drinks and innovative new recipes. Craft cocktails were in vogue again, and their popularity has only increased since.

The digital age and the rise of social media have also had an effect on the cocktail world. While a well-presented drink was always a goal of these craft cocktail bars, the ubiquity of the cellphone camera and the popularity of Facebook and Instagram have driven this to a new extreme. Like the transition from shared punch to single-serving cocktails, it marks a change in the way we view drinking. In a way, it's a movement toward an even more individualistic mindset—every drink needs to be unique and reflect a personal aesthetic.

But I prefer to see it as a regression back to something more communal, a virtual punch bowl around which we can all gather.

Another recent movement has been the growing popularity of nonalcoholic cocktails, or "mocktails." At many bars and restaurants, mixologists are crafting these alcohol-free drinks with the same care and skill they use for cocktails, with creative ingredients like shrubs, house-made sodas, and even nonalcoholic distilled spirits. For the home bartender, it's always a good idea to have some mocktail recipes up your sleeve. There are a number of reasons why you or a guest may not be drinking, and it's always nice to be able to serve something that feels as special as a cocktail.

When making a proper drink, whether it's a cocktail or a mocktail, fresh ingredients are critical. We have officially escaped the era of sugary mixers and neon-red cherries. Good bars use fresh-squeezed juice, house-made syrups, and fresh herbs, and home bartenders should do the same. And while you can probably run to the grocery store for what you need, the best way to have fresh ingredients available is to grow them yourself. There's nothing better than being able to run out into your garden to pick your own herbs and produce. There's also no better way to dress up your go-to recipes than adding fresh, seasonal flavors.

In this book, you'll find twenty-five classic cocktail recipes that every home bartender should know how to make, along with twenty-five original variations on these that take advantage of fresh ingredients and garnishes from your garden. I hope that this book not only introduces you to some new favorites, but also shows you how easy it is to create your own drinks with flavors you love.

Whether you've never made anything more complicated than a Jack & Coke or you're already posting elaborately garnished cocktails to Instagram, I think there will be something in this book for you. But I wrote it with beginners in mind. It's the book I wish I had when I started making drinks and when I first planted my garden. I hope that it will help you do a little magic of your own as well.

Making Cocktails

The Bottles You'll Need

Building a home bar can be an intimidating process. Alcohol is expensive, and a trip to the liquor store can quickly become overwhelming. There are a lot of different products out there, and it's hard to know which ones you'll actually like and use. So, in this section I am going to introduce the bottles that I recommend for the cocktails in this book and try to make it easier for you to pick out which ones you want to buy if you don't already have them. While the recipes that follow do use all of the major categories of spirit (Cognac, gin, rum, tequila, vodka, and whiskey), I've tried very hard to limit the number of liqueurs and modifiers. I know that these can get expensive, and seeing too many unfamiliar ingredients in a recipe can be intimidating. You can make every cocktail in this book with fourteen bottles:

- Cognac
- Gin
- Rum
- Tequila
- Vodka
- Whiskey
- Dry vermouth

- Sweet vermouth
- Campari
- Elderflower liqueur
- Maraschino liqueur
- Orange liqueur
- Angostura bitters
- Orange bitters

I'll go into detail on each of these below. I chose them because they are all ingredients that will get a lot of use in your home bar if you get into making cocktails. They show up in a number of classic and popular drinks. I promise that, if you buy other cocktail recipe books, you'll find a lot more recipes that you can make with just these fourteen bottles. They are the perfect basis for your home bar.

That said, don't be afraid to make substitutions if you need to! If one of the gin cocktails sounds really good and all you have is vodka or tequila, try it out. Sometimes it may not quite work, but sometimes it will be delicious. There are a lot of famous drinks like the Oaxacan Old Fashioned or Kingston Negroni that are made by switching out the base spirit in a classic (whiskey for tequila in the Old Fashioned and gin for rum in the Negroni, respectively). Experimenting can be fun and helps build confidence in your mixology skills.

After I discuss each of these categories, I'll recommend some bottles for you to buy. I've tried to choose well-regarded brands at a reasonable price point. For categories where several bottles are listed, I've put my top recommendation in bold. These selections are my personal opinion and some people will certainly disagree with them. Choosing the right spirits for your bar is always going to be a matter of personal preference. I recommend several in each category, not to overwhelm you with choices or to suggest that you need them all, but to offer alternatives if your local liquor store doesn't stock everything. All the brands I've listed here will make a fantastic cocktail, and so will whatever you already have at home!

Cognac

Cognac is a type of **brandy**, which is a spirit made by distilling fruit juice. Grape brandies are most common but there are also apple brandies, pear brandies, and other varieties. Cognac is a brandy distilled from grapes that is made in the Cognac region of France.

If you're not already into Cognac cocktails, then the mention of this spirit probably brings to mind either an old man in a smoking jacket sipping from a snifter in his study or Kanye West drinking straight from a bottle of Hennessy at the MTV Movie Awards. Both are accurate, I suppose, but neither really captures the spirit of Cognac in the craft cocktail world. Before the phylloxera blight devastated France's vineyards in the 1870s, Cognac was the spirit of choice in cocktails. The two classic Cognac cocktails in this book, the Brandy Crusta (pg. 87) and the Sidecar (pg. 91), are elegant and delicious recipes that translate beautifully to the modern bar. And

Cognac mixes well with a variety of garden flavors, like apple, pear, rosemary, and tarragon. It definitely deserves a place in your home bar.

Recommended bottles:

Hine H VSOP, **Pierre Ferrand 1840**

Gin

Gin is simply a distilled grain spirit flavored with juniper and other botanicals. Every gin is going to taste a bit different depending on what botanicals are used, but the flavor of juniper is what ties them all together. The main category you should know is **London Dry**, which is characterized by a lot of juniper flavor and is, well, dry (as in not sweet—especially in comparison to older styles like Old Tom and Genever). Examples include the classic gins you certainly know: Beefeater, Gordon's, Tanqueray, and Bombay Sapphire. You can't go wrong with a London Dry for your cocktails. But if you find the juniper flavor overpowering or want to try something different, there is an emerging category often called **Contemporary** or **New Western** gin that puts less focus on juniper, letting other botanicals shine through more. If a gin bottle doesn't specify London Dry, it probably falls into this category.

Recommended bottles:

London Dry: Beefeater, Ford's, **Sipsmith**

New Western: Aviation, Barr Hill, Empress 1908, Hendrick's

Rum

Rum is a spirit made by distilling sugarcane products, usually molasses. It can seem like a very simple category at first glance (there's white and dark rum) but gets incredibly complicated as you really start learning about it. Rums differ in flavor depending on sugar source, still type, and aging method. Because these things are usually consistent within the country or island where the rum is made, it is often categorized by region.

Because I don't want to get into the nitty-gritty of rum production here, I will keep things simple and specify either a white rum or an aged rum. **White rum** has a light flavor and is either unaged or briefly aged and then charcoal filtered. **Aged rums** have been barrel-aged and have a richer flavor.

Recommended bottles:

White rum: Bacardi Superior, Caña Brava, **Plantation 3-Star**[1]

Aged rum: **Appleton Estate Signature Blend**, Bacardi Reserva Ocho, Plantation Original Dark

Tequila

Tequila is a distilled spirit made in Mexico from blue agave (*Agave tequiliana*). The leaves and thorns of the plant are stripped away, and the heart of the agave plant is roasted and crushed to collect its juice. This liquid is fermented,

---

[1] Plantation announced in June 2020 that it will be changing its name. The new name has not been released at the time of writing.

distilled, and aged. There are three main categories you should know:

Tequila Blanco or Plata is aged for less than two months and is usually clear. It has a clean agave flavor.

Tequila Reposado is aged for between two months to one year and has a faint golden color. It begins to take on some of the rich vanilla flavor of an aged tequila.

Tequila Añejo is aged for one to three years and has a deep color and strong oak and vanilla notes. It's often sipped like a whiskey. There is also an Extra-Añejo category that refers to tequila aged for more than three years.

The tequila drinks in this book all call for a blanco tequila, though a reposado would also work.

Recommended bottles:

Espolon Blanco, Don Julio Blanco, Fortaleza Blanco, Tapatio Blanco

Whiskey

Whiskey is a spirit distilled from a fermented grain mash and then aged. This category includes bourbon, rye, and Scotch, among others. The difference in flavor between these types and between individual whiskeys comes from a number of sources, including what grains are used (the **mash bill**), the length of aging, and the type of barrel used. The recipes in this book can be made with either a bourbon or a rye whiskey, depending on your preference. I don't recommend using Scotch or Irish whiskey, as their flavor can be quite different.

Bourbon is whiskey that is made in the United States with at least 51 percent corn and aged in new charred oak barrels. The high proportion of corn makes bourbon sweeter and more full-bodied than other whiskeys.

Rye has the same requirements as bourbon, but must be at least 51 percent rye, which makes it fruitier and spicier than bourbon.

Scotch is made in Scotland, primarily from malted barley. **Malting** is a process of drying the grain, and bourbon and rye often contain some percentage of malted barley as well. While some Scotches can be very delicate in flavor, others have bold flavors, like the smoky whiskies from Islay that use peat fires to malt their grains.

Irish whiskey has very similar requirements and terminology to Scotch but is made in Ireland.

If you think you're not a whiskey drinker, try the Whiskey Smash on pg. 193. It might just convert you!

> **Recommended bottles:**
>
> Rye: **Old Overholt**, Redemption, Rittenhouse
>
> Bourbon: **Buffalo Trace**, Elijah Craig Small Batch, Knob Creek

Vodka

Vodka is essentially a colorless, flavorless distilled spirit. It is usually made from fermented grain, but it can be made from just about anything—potatoes, honey, fruit, or even milk (check out Black Cow Vodka from the UK).

Vodka gets a bad rap among cocktail nerds who prefer more flavorful spirits, but it has its staunch defenders, and it remains the most popular spirit in the United States. It's a fixture in many classics like the Bloody Mary (pg. 163), Cosmopolitan (pg. 167), and Moscow Mule (pg. 171), and is a popular substitute in some gin cocktails, like the Gimlet (pg. 109) and Martini (pg. 117).

Recommended bottles:

Ketel One, **Reyka**, Tito's Handmade Vodka

Vermouth

Vermouth is fortified and aromatized wine. **Fortified** means that a high-proof spirit, usually brandy, has been added to increase its alcohol content and extend its shelf life. **Aromatized** means that it has been flavored with botanicals—in the case of vermouth, this usually includes wormwood. While quite enjoyable on its own over ice as an aperitif, vermouth is most often used as a modifier in cocktails like the Martini and Manhattan.

Despite the extra boost in ABV (alcohol by volume), vermouth is still wine, and it can oxidize and go bad. Which brings me to the most important instruction you will receive in this section of the book: *refrigerate your vermouth.* If you've only ever had a Manhattan made with room-temperature sweet vermouth from a bottle in the back of a liquor cabinet that's still covered in a fine coating of dust, then you probably think you don't like Manhattans. Vermouth goes bad. But if you store it in the refrigerator, it will stay good for weeks or months, and it will taste *much* better.

The three main types of vermouth are **dry**, **sweet**, and **blanc or bianco**. Most are made in France and Italy, and French and Italian vermouths differ in style and flavor. It can get a bit confusing, as sometimes the term "French vermouth" is used to refer to dry vermouth and "Italian vermouth" to designate sweet vermouth. This is mostly historical—today, most brands in both countries produce all three types.

Dry: This vermouth, used in the Martini (pg. 117), is colorless to pale yellow and not very sweet.

Sweet: The vermouth of choice for the Manhattan (pg. 177) and Negroni (pg. 121) is dark red in color and contains more sugar than dry vermouth. The Italian styles will be sweeter and more full-flavored than the French varieties.

Blanc: Though it is called for less often, blanc or bianco vermouth is worth knowing as well. This is a sweeter white vermouth. I prefer it in the Strawberry Blonde (pg. 122), but sweet vermouth will also work.

> *Recommended bottles:*

Dry vermouth: Dolin Dry

Sweet vermouth: **Carpano Antica Formula**, Cocchi Vermouth di Torino, Dolin Rouge

Blanc vermouth: Dolin Blanc

Campari

Campari, an Italian bitter liqueur with a distinct bright red color, is made by infusing alcohol with a mixture of herbs and botanicals. The precise recipe has been kept carefully secret.

It has been produced since 1860 and is a fixture in countless classic cocktails. It is popularly enjoyed as an aperitif before a meal, mixed with soda and/or vermouth, or stirred into a Negroni (pg. 121).

Because it is so bitter, you may find Campari to be an acquired taste. I have always quite enjoyed it mixed with soda and an orange slice, but I hated the Negroni the first time I tried it. If Campari feels like too much for you, ease your way in with some Aperol, a similar liqueur that is a bit sweeter and less bitter. If you enjoy an Aperol Spritz, you're well on your way to becoming a Campari drinker one day.

Recommended bottle:

There's only one Campari!

Elderflower Liqueur

Unlike the other products on this list, which have been made and used for decades or even centuries, elderflower liqueur is an extremely new addition to the cocktail world. It was introduced in 2007, when Robert Cooper brought St. Germain elderflower liqueur to the market. It was an instant phenomenon. The reason for its popularity is not difficult to see. It is a delicious liqueur with a crowd-pleasing floral flavor that shines in both simple and complex cocktails. Several more brands have popped up since 2007, including the budget-friendly St. Elder, but St. Germain still reigns supreme in this category.

Recommended bottle:

St. Germain

Maraschino Liqueur

Maraschino liqueur is made by distilling sour marasca cherries, which are native to the Dalmatian coast. Though the cherries are sour, the liqueur itself is very sweet, and can lend a complex, candied-cherry flavor to a cocktail. In this book, you'll only find maraschino liqueur in the recipes for the Brandy Crusta (pg. 87) and Upstate (pg. 178), but it also appears in classic cocktails such as the Aviation, Last Word, Red Hook, and Hemingway Daiquiri. As such, it's a worthy addition to any home bar.

Recommended bottle:

Luxardo Maraschino Liqueur

Orange Liqueur

Orange liqueur is a common ingredient in many classic cocktails, from the Margarita to the Mai Tai. It is made by including orange peels in the distillation of a spirit. The category is often divided into **triple sec** (which includes the very popular Cointreau) and **orange Curaçao** (also called dry Curaçao). But the difference between these categories is very fuzzy. Grand Marnier, one of the more popular orange liqueurs, doesn't fit readily into either of them. Triple sec is usually colorless and made with a neutral spirit, whereas Curaçao is made with rum or brandy. I find that Curaçao has deeper, spicier notes, while triple sec has a brighter, candied-orange-peel flavor.

In this book, I will specify either orange Curaçao or triple sec when orange liqueur is called for. You can substitute one for the other if needed.

Recommended bottles:

Curaçao: Pierre Ferrand Dry Curaçao

Triple Sec: Cointreau

Bitters

Bitters are highly concentrated alcoholic solutions flavored with herbs, spices, and bittering agents. They were originally formulated as medicinal tonics but were co-opted for use in cocktails. In fact, the original definition of a "cocktail" is spirit, sugar, water, and bitters. Because they are so intensely flavored, bitters are usually only added to drinks in small dashes. But they can have a huge effect on a drink. Bitters help to bind the other ingredients of a drink together and bring out certain flavors. They're like the salt and pepper of the cocktail world.

Today, there are dozens of brands of craft bitters and hundreds if not thousands of flavors and varieties. Despite this, there is one type of bitters you will find in cocktail recipes more commonly than any other: Angostura. This Venezuelan-made bitters, with its iconic oversized label, has been around since 1824 and is a crucial element to many classic and modern cocktails. It's deep red in color and has strong flavors of clove, cinnamon, and other spices.

The next most common variety of bitters that also appears in some of the recipes in this book is orange bitters, in which orange peel is the main flavoring agent.

Recommended bottles:

Angostura Bitters, Fee's Orange Bitters, Regan's Orange Bitters

Nonalcoholic Ingredients

Syrups

Learning to make simple syrup is a critical skill for a home bartender. Simple syrup is probably the most common ingredient that is called for in cocktails, and it is so easy to make. Once you know how, you're going to be able to make all sorts of variations, from honey syrup to flavored syrups.

Simple syrup is just equal parts water and sugar. Combine them in a small saucepan and set it on your stove on medium-low heat. As it begins to simmer, stir until the sugar is dissolved. Let it cool, and you're ready to use it. You can also cheat by using the microwave in a pinch. Simple syrup can be stored in the fridge for several weeks. To prolong its life, add a dash of a high-proof spirit to the bottle.

You can build on this basic template to make all sorts of other syrups. Adding spices or herbs and then straining them out will give you a flavored syrup. So will muddling in fruit or using a fruit juice instead of water. You can also substitute honey, agave nectar, or other sweeteners for the sugar. In the recipes that follow, I'll provide exact instructions on how to make the various flavored syrups in the drinks.

Mixers

Club soda: Club soda is carbonated water to which a small amount of potassium and/or sodium salts has been added. It's essentially interchangeable with other types of carbonated water, such as seltzer, soda water, and sparkling or mineral water. While you can spend quite a lot on some mineral waters, I don't see much of a difference between different types and brands myself.

Ginger beer: An essential ingredient in the Moscow Mule (pg. 171), ginger beer is a sweetened and carbonated ginger soda that is spicier and fuller-flavored than ginger ale. Good brands to look out for are Q Mixers, Fever Tree, Barritt's, and Goslings.

Sparkling wine: While sparkling wine is not technically a mixer since it contains alcohol, it can essentially be treated like one. When choosing a wine to mix with, look for something that isn't too sweet. Prosecco is usually a good choice, and there are several high-quality options in the eight-to-fifteen-dollar range that are perfect for use as mixers. Mionetto and La Marca are two that are widely available and go well in cocktails.

Tonic water: Not to be confused with the other carbonated waters mentioned above, tonic water is flavored with quinine, which makes it very bitter. It was first created as a more palatable way to take quinine powder for the prevention of malaria. Now it is enjoyed for its bitter, refreshing flavor—usually with gin. Use Schweppes or Canada Dry if you like it, but I do encourage you to try a premium tonic like those made by Fever Tree. I find the difference in flavor and quality to be well worth the extra cost.

Tools

Crafting cocktails is best done with a very specific set of tools. In this section, I'll introduce the basic tools I think you should have to make cocktails at home. If you don't have everything on this list, don't let that stop you from making drinks. There are lots of things you probably do have in your kitchen that will work just fine at first, and once you feel more confident making cocktails at home, you can invest in some proper tools. For that, I highly recommend Cocktail Kingdom (cocktailkingdom. com). They supply both professionals and home bartenders, and their products are extremely high quality. There are plenty of cheaper sets out there as well—check online for dozens of options—and that's a great way to start out.

Shaker

There are two kinds of cocktail shakers, the **cobbler shaker** and the **Boston shaker**. A cobbler shaker has three pieces: a tin, a lid with a built-in strainer, and a small cap. In contrast, a Boston shaker is simply two tins, or a tin and a glass, that fit together (see the shaker in the photo on the previous page). The cobbler shaker is popular among home bartenders, but if you start paying attention, you'll see that most professional bartenders use Boston shakers. They're less likely to get stuck, and they give the drink more distance to travel with each shake. I recommend a tin-on-tin Boston shaker, as glass can break more easily.

If you're only going to buy one tool for your bar, get a shaker. You could maybe use a Mason jar in a pinch, but otherwise there aren't many good substitutes.

Mixing Glass

Some cocktails are shaken, but others are stirred (see the Techniques section below for more information on which is which). For stirred cocktails, bartenders use a mixing glass—a large glass with a spout. You can substitute the tin of your shaker if you don't have one.

Strainer

If you use a Boston shaker or a mixing glass, you'll need a strainer to keep the ice from falling into your glass when you pour out your chilled drink. The most popular kind is a **Hawthorne strainer**, which has a spring and prongs to help it fit just right over the rim of your tin. You may also come across the **julep strainer**, which looks like a large spoon with holes in it. These are generally used for stirred drinks, but it's mostly a stylistic choice, as the Hawthorne will work fine for either.

I also highly recommend getting a small fine mesh strainer, like the kind sometimes used for pouring tea. Small ice chips and bits of lemon pulp will slip through the holes on a Hawthorne or julep strainer, so it's often recommended that you "double-strain" or "fine strain" your drink by placing this over the glass to catch any smaller undesirable bits.

Barspoon

A barspoon is a long, spiraled spoon that is used to stir a cocktail. It's also a unit of measurement—one barspoon is equal to one teaspoon. If you don't have a barspoon, use something long and thin to stir your drink, like a chopstick.

Jigger

Precise measurements are critical to making a good cocktail. The tool that bartenders use for this purpose is a jigger. This is a double-sided tool with two small cups used for measuring. The two sides hold different amounts of liquid. Jiggers come in a variety of combinations, such as 1½ ounces and 1 ounce, 1 ounce and ½ ounce, or ¾ ounce and ½ ounce. You can use any size, but a jigger that has at least a 1-ounce measure will be most practical. Look for one with lines on the interior for smaller measurements.

If you don't have a jigger, a set of teaspoons can work. One tablespoon is equal to half an ounce.

Lewis Bag and Mallet

If you've ever had a particularly frustrating day, I highly recommend making a cocktail with crushed ice so that you can take out your Lewis bag and mallet. This cloth bag and wooden hammer are used for crushing ice. You simply fill the bag with ice, close the flap, and crush the ice with the hammer until it is at your desired consistency. It's particularly convenient because ice doesn't stick to the inside of the bag the way it might to a kitchen towel or other cloth. But if you don't have one, try an unsealed Ziplock bag wrapped in a kitchen towel, and any heavy, broad kitchen tool.

Muddler

A muddler is used for gently crushing herbs, berries, fruits, or vegetables to incorporate their flavor into a cocktail. Look for a wooden muddler with a smooth base rather than one with teeth—the idea is to gently bruise your herbs, not tear them apart, which will release more bitterness into the drink.

Techniques

Reading Recipes

If you start to pay attention to cocktail recipes, you'll see that the ingredients are usually listed in a standard order. They start with the things that contain alcohol, from highest volume to smallest. Then come the remaining ingredients, again in order of volume. Things that are meant to be added directly to the glass, like carbonated mixers, or that are incorporated in a special way, like muddled herbs or rinses, are usually placed last. Bartenders will often add ingredients from the bottom up, so that the most expensive ones—the spirits—will come last. This is in case they make a mistake and have to toss out the drink. You may want to do the same.

Measuring

Proper measuring of ingredients is essential to achieving a good cocktail. When using your jigger to measure, fill it up completely. You want the liquid to be perfectly level with the top of the jigger, not dipping in or bulging out.

Also, keep in mind that all measuring implements are going to vary a little. The cheaper your bar tools, the more likely this is. You may want to check the measurements on your jiggers with a set of teaspoons. Or be sure to use the same jigger for all measurements in a cocktail so that, if it's off, it's off consistently.

In the US, almost all amounts in cocktail recipes are provided in ounces. The common exceptions are a **barspoon**, which is equal to one teaspoon, and a **dash**, which is equal to ⅛ teaspoon. Bitters generally come with dasher tops, so that you can invert the bottle and give it a quick shake to dispense one dash. In the Bloody Mary (pg. 163), I also use a **pinch** for the spices—this is also ⅛ teaspoon and is more commonly used for dry ingredients.

You may also see a recipe that simply says "club soda, to top" or something along those lines. This can be confusing, since everything else in the recipe is so precise! But as long as you're not using a massive glass, you can just pour the mixer until the glass is full. See the Glassware section below for standard glass sizes.

Shaken or Stirred?

There are a few simple mixology rules that you can learn quickly and remember forever, and here is one: *shake a drink* when it contains juice, and stir it when it contains only spirits. Shaking helps to emulsify and combine many ingredients of different densities. Stirring, on the other hand, will easily mix alcoholic ingredients, and helps them maintain the smooth, dense texture that is desirable in a Martini or Manhattan. This means, of course, that James Bond commits a terrible faux pas when he orders his Martini shaken!

The goal with shaking or stirring is to get your drink to the proper temperature and dilution while combining your ingredients. And you do want to dilute your drink—a little. Your spirits are full of long chains of organic compounds called **esters.** Esters give all food and drinks their flavor. Water slices them open, releasing smaller, more volatile esters, enhancing the flavor of your drink.

Shaking

When shaking a drink, begin by combining all your ingredients in your shaker. Pay attention to the recipe you're using, as some things may be added later, like carbonated mixers. And don't fill the shaker with ice until right before you're ready to shake. This gives you plenty of time to properly measure everything without worrying about the ice melting and the cocktail getting watery.

When you're ready to shake, fill the shaking tin with ice and seal up your shaker. With a cobbler shaker, be careful to put the lid on straight, or it may get stuck. However, with a Boston shaker, you want the tins to join at a slant. Fit them together at an angle and rap the one on top with the heel of your hand to form a seal.

You should shake your drink for about ten seconds, until the outside of the shaker becomes frosty and you can feel that the ice has begun to break up a little. When you're finished, either uncap your cobbler shaker or separate the tins of your Boston shaker (try giving the side of the larger tin a firm whack with the heel of your hand about a quarter of the way around from where the tins meet). Then strain your cocktail into your glass using the built-in strainer of the cobbler shaker or a Hawthorne strainer fitted into your Boston shaker. If the drink contains muddled herbs or pulpy juice, fit a fine mesh strainer over the glass before you pour.

Stirring

Stirring is a simpler process than shaking. Simply combine your ingredients in the mixing glass, add ice, and stir steadily for fifteen to twenty seconds. When the drink is ready, use a strainer to pour it into your glass.

Note that if your drink is served on the rocks, you want to fill your glass with ice before straining in your drink. This is true regardless of whether the cocktail is shaken or stirred.

Building

Occasionally, you will see the instruction to **build** a drink in a glass. This means pouring all the ingredients directly into the glass you will serve it in, rather than shaking or stirring first. This is sometimes done with spirit-forward drinks like the Old Fashioned (pg. 189) or Negroni (pg. 121) (especially if you're just quickly making one at home) or with simple drinks like the Gin & Tonic (pg. 113).

Rimming a Glass

Rimming a glass with salt or sugar adds a dynamic element of texture and flavor to every sip. To rim a glass, gently rub a slice of lemon or lime (whichever you're using in the drink) around the edge of the glass to wet it, then lower the glass into a shallow dish full of salt or sugar, wiggling it slightly until it is coated. I'll often do a half-rim so that I can choose whether to sip from the rimmed portion or not. You can also make your rims more dramatic (though less practical) by running them further down the glass, as I did for the Paloma (pg. 157) and Thai Bloody Mary (pg. 164).

Muddling

Muddling is all about releasing the flavors of ingredients like herbs or citrus into your drink. To muddle, put the listed ingredients into the bottom of your shaking tin or mixing glass and gently press on them with your muddler. When muddling herbs, you simply want to press out the aromatic oils that will flavor the drink, not tear them or mash them to a pulp, which can release bitterness. Even muddling fruit to release its juices doesn't require too much force.

Ice

Ice for Shaking and Stirring

As I mentioned above, the purpose of shaking or stirring a drink is to chill and dilute it. The size of the ice you use will affect how this happens. Ice with more surface area will melt faster, and if you use cubes that are too small, they may over-dilute your cocktail. Conversely, cubes that are too large will take longer to do the job. Most bars use cubes that are about one-inch square. The ice from your freezer ice machine or ice trays will be just fine. You should not use crushed or pebble ice unless it is specifically called for, as it will dilute your cocktail too much.

Ice for Serving

The ice that goes in your final cocktail can take a few different forms. The most common way to serve a drink with ice is "on the rocks," or, as I usually write in these recipes, "strain into a rocks glass filled with ice." Your normal freezer ice is fine for this. If you want to dress things up a bit and emulate your favorite craft cocktail bar, you can buy a mold that makes large cubes or spheres. These are generally used for spirit-forward drinks that should dilute more slowly, like an Old Fashioned (pg. 189) or a Negroni (pg. 121). And finally, if a recipe calls for crushed ice, whip out that Lewis bag and mallet and go to town. If your freezer will do it for you, that works too.

Speaking of your favorite craft cocktail bar...you may be wondering how they get such clear ice. You may even have bought a special tray to make cubes or spheres that shows perfect, crystal-clear ice on the packaging, only to be disappointed by white and cloudy cubes when you make your own at home. Achieving clear ice is actually a complicated process. But it can be done if you really want to, and it does feel quite fancy to have clear ice at home.

Water freezes directionally, from the top to the bottom. As it freezes, the impurities in the water that cause it to be cloudy will sink and freeze last. This is particularly true when the freezing happens slowly, giving the impurities plenty of time to sink. So, the way to get clear ice is to slow the freezing process with an insulated container, and to take advantage of this directional freezing. This can be as simple as filling a Styrofoam ice chest with water and putting it in your freezer— the top of the resulting block will be clear, and you can saw it off and cut it into cubes. But if you want clear ice on a regular basis, the easiest way is to buy a specialized system for it. I own an ice chest made by Wintersmiths that produces clear cubes and spheres. It's pricey, but it works wonderfully. There are also a couple of local companies near me that sell clear ice to bars and consumers—search for similar retailers near you and see if you can buy a bag of clear spheres for your next cocktail party.

Glassware

There's no denying that drinks seem to taste better when you sip them out of fancy glassware. And there may actually be some truth to it. Tall, skinny glasses help maintain carbonation. Stemmed glasses keep your drink away from your hand, which would warm it up. Snifters funnel the aroma of the drink to your nose. There's usually a reason why a specific glass is recommended for a cocktail.

What follows is by no means an exhaustive reference on cocktail glassware, which could get quite lengthy—these are just the glasses called for in this book.

Glasses for Iced Drinks

Collins or Highball Glass: These are tall, narrow glasses often used for fizzy cocktails, usually holding ten to twelve ounces.

Rocks, Old Fashioned, or Lowball Glass: Also called a tumbler, this is a low glass that holds around eight ounces and is used for drinks served on ice.

Stemmed Glassware

Cocktail or Martini Glass: While this iconic glass has fallen out of favor in most craft cocktail bars, replaced by the more practical coupe glass, it's still a fine choice for cocktails at home if you like the shape. Use it anywhere a coupe is called for in this book.

Coupe Glass: The bowl-shaped coupe glass, originally created to serve champagne, has become a symbol of the cocktail's golden age and has experienced a revival in recent years. It's the go-to glass for cocktails served without ice. A good size is six ounces.

Champagne Flute: The tall, elegant Champagne flute is specially shaped to help the wine retain its carbonation. It's a common choice for cocktails containing sparkling wine that are served with no ice.

Specialized Glasses

Julep Cup: A Mint Julep (pg. 181) is traditionally served in a silver cup that becomes frosty from the crushed ice inside.

Moscow Mule Mug: The creators of the Moscow Mule (pg. 171) chose to serve them in this distinctive copper mug so that everyone could recognize the drink. Like a julep cup, the mug will become nicely frosted when filled with a cold beverage.

How (and Why) to Garnish

Every bartender has their own philosophy on garnishes, from the minimalists, who prefer to twist an orange peel over the glass and discard it, to the cocktail Instagrammers, who tend to go a little over the top. I obviously fall into the latter category. I think there are three things a garnish can add to a cocktail: flavor, aroma, and visual appeal. And it doesn't have to do all three. While all of the flowers you see in this book are edible, I wouldn't really recommend chowing down on most of them as you sip your drink. They're there to make the cocktail look good. We drink with our eyes first, and anything that increases your overall enjoyment of a cocktail is a valid addition. But don't let the fact that you don't have a garden full of edible flowers (yet—see the next section!) stop you from making drinks. There are a lot of ways to garnish a cocktail, and some affect the flavor of the drink more than others.

Citrus Twists

Citrus twists are the most common cocktail garnish, and, when they are done correctly, they can add an immense amount of flavor and aroma to a drink. I used to think that garnishing something "with a twist" meant adding a thin, curly piece of lemon peel, but this isn't actually the case. The *twist* refers not to the shape of the peel, but to the action you perform as you garnish the drink.

When doing a twist, use a sharp knife or Y-peeler to remove a large (one-by-two-inch) piece of peel from your citrus fruit. You want to cut shallowly, so that you get as little of the white pith as possible. Take this piece of peel and, with the outside of the peel facing your drink, twist or squeeze it gently to release its oils over the drink. In the right light, you should be able to see the oils squirt out and hit the surface of your cocktail. Next, rub the outside of the peel along the rim of the glass. Then either drop it into the drink or discard it, whatever your preference. Try taking a sip of the drink before and after doing this—you'll notice a huge difference.

If you still want the look of a curled twist, try curling up the piece of peel before adding it to your drink. You may need to hold it in place for a while before it will stay curled. Coiling it around your barspoon is one way to help shape it.

Fresh Herbs

Fresh herbs are another garnish that can really affect the flavor of a drink. They're usually tucked into ice, floated on the surface, or clipped onto the side of a drink. They primarily add aroma to the cocktail, which in turn changes the way you taste the drink. One of the most common herbal garnishes is a mint bouquet.

Edible Flowers

Edible flowers are usually added to cocktails for their visual appeal, but there are a few exceptions. The flowers of herbs are usually quite aromatic and will behave much like the herbs themselves when garnishing a cocktail—examples include mint, basil, and thyme flowers. Extremely fragrant blossoms like roses, jasmine, or honeysuckle will also have an effect on flavor.

In the next section of the book, I will list some edible flowers and tell you how to plant your own cocktail garden. But you can also purchase edible flowers from online retailers like Gourmet Sweet Botanicals (gourmetsweetbotanicals.com), which supplied some of the garnishes for this book.

Other Garnishes

When it comes to cocktail garnishes, the possibilities are endless! But there are a number of other common garnishes you may encounter.

Brandied cherries: A neon-red maraschino cherry may do in a pinch, but do yourself a favor and pick up a jar of Luxardo brandied cherries. This is one of the few garnishes that you're actually expected to eat, so it's a good place to buy a quality product.

Cucumber ribbon: Cucumber can make a lovely, dramatic garnish when sliced very thin using a Y-peeler, mandoline, or sharp knife. I used cucumber ribbons to garnish the Cucumber Collins (pg. 130) and the Gin & Tonic (pg. 113).

Dried citrus: Dried citrus slices can be more photogenic than their fresh counterparts. They also last longer and are more likely to float in your drink. To make your own, slice citrus very thin and place in a food dehydrator or an oven set to 200°F until they dry out. You can also purchase them from retailers such as the Cocktail Garnish (thecocktailgarnish.com).

Olives: Commonly called for to garnish a Martini (pg. 117) and a Bloody Mary (pg. 163), olives are another garnish that you're actually going to be snacking on as you drink your cocktail, so it's good to find some quality ones. I love crisp, bright green Castelvetrano olives in my drinks.

Spices: Dried spices like cloves, star anise, or cinnamon sticks can be floated in your drink to impart some of their scent and flavor to the cocktail. I particularly love adding these to whiskey cocktails like the Fall Fashioned (pg. 190), but they can be surprising additions to something like a Gin & Tonic as well (pg. 113).

Growing a Cocktail Garden

If you want fresh ingredients and beautiful garnishes for your cocktails, the best way is to grow them yourself. If this sounds like an intimidating prospect, this book is for you! Keeping plants alive and healthy isn't nearly as difficult as you might think. You don't even need a yard. Until very recently, I grew all of my plants in pots on a small urban balcony. In some ways this is easier—you can control your soil and drainage, and you don't have to worry about weeds.

If you're just starting out, choose plants that are easy to grow and that you'll get a lot of use out of. I recommend mint, rosemary, viola, and dianthus. They're all easy to find at local nurseries at the beginning of spring, are hard to kill, and look beautiful as garnishes.

When you start planning your garden, it's easy to get very intimidated by the specific requirements of each type of plant. Each one has its own preferences for soil, temperature, water, and sun. But as you read more and start growing things yourself, you'll see that there are a lot more commonalities in what plants require than differences. You'll also see that many plants are quite forgiving. I really don't want to overwhelm you with small details in this section. I want to equip you with the basic, practical knowledge you need to get out there and grow your garden. So, I'm going to cheat a little and tell you some general things that should work for most if not all of your plants. If you have success and want to get more into the details of what each plant likes best to really thrive, there are plenty of more comprehensive resources out there to help you take the next step.

When to Plant

Early spring is the best time to plant your garden. You generally want to wait until after the last frost to plant anything outside, since the cold could kill your young plants. There are online tools that will let you look up the average date of the last frost in your specific area. Try almanac.com/gardening/frostdates.

Starting from Seeds

There are a few reasons why you may want to start some of your plants from seeds. For one thing, it's much cheaper to buy seeds than established plants. It's also very easy to order seeds online, which is helpful if there is a plant you want that your local nursery doesn't carry. You may also be able to save seeds from your annuals to plant the following year so that you don't have to buy them again. It's also very satisfying to raise a big, healthy plant from a tiny seed!

The downsides of seeds are that they obviously take longer to grow, and they're not always a sure thing. There are many reasons seedlings may fail to thrive, whereas established plants are more robust. I generally opt for established plants when possible. But there are many plants in my garden that I have grown from seed.

If you want to be ready to plant your seedlings outdoors once the last frost has passed, you can start your seeds inside several weeks beforehand. Your local nursery may sell special trays for this purpose that have small pots, a base to catch water, and a lid to help keep moisture in. You can also buy seed starter mix, which is formulated to hold in the correct amount of moisture for germination.

Plant your seeds according to the individual package directions. The odds are that they will not all sprout, so put several in each of your pots. You want to keep the soil moist but not too wet, especially once the seeds have germinated. Make sure they get plenty of sun. Once the seedlings have two sets of leaves, choose the healthiest one in each pot and cut or pull out the rest. When they have three to four true leaves, they're ready to move outside. To really set them up for success, you can do what's called **hardening them off**, a process by which you bring them outside for increasing amounts of time over the course of about a week. It helps ease the transition to life outside.

Starting from Cuttings

If you have friends with their own gardens, starting new plants from cuttings is a great compromise between budget-friendly seeds and established plants. Not all plants can be propagated from cuttings, but it will work for a surprising number of species.

Some plants do best with a softwood cutting that takes a fresh, younger stem, whereas others should be propagated from a hardwood cutting from an older part of the plant. Do some research on the species you're trying to propagate to determine what will work best.

Take your cutting from a healthy plant. Choose a stem and make a clean cut a few inches down the stem, being sure to include a few sets of leaves. Remove the lowest leaves that will end up in the potting medium or water. As for what to use, there are several options. Propagating in water can work for many plants and is great because you can see the roots as they form. But a proper medium like perlite, vermiculite, sand, or a mixture of these is preferable. Potting soil or seed starter is another option that may work. If you really want to set yourself up for success, get some rooting hormone, which is a powder that helps encourage the cutting to produce roots.

Plant your cutting in the medium, or place it in the water, just deep enough to keep it upright. It should remain moist, so check the soil frequently and water as needed. Many people recommend placing a plastic bag over it to create a mini-greenhouse. Place the cutting somewhere warm where it will get lots of light. If all goes well, the plant should produce roots in a month or two and be ready to transplant into the garden.

Try very gently tugging on the stem to see if it has rooted itself in the medium. As with seeds, hardening off by slowly transitioning the plant outdoors is recommended.

Choosing Where to Plant

Sun: I'll go into the exact sunlight needs of each type of plant below, but you'll find that most plants need as much sun as you can give them. If you have any sections of your garden that are particularly shady, look for plants that can still thrive in shade, like begonia or fuchsia.

Soil type: The US Department of Agriculture classifies all soils by their makeup of sand, silt, and clay. Soil with a certain combination of these three types is called loam. Loamy soil is the best for plants. It has the ideal texture to retain water and nutrients while allowing oxygen flow to the roots of the plant. But the odds are that you don't have loamy soil, and that's ok. You can improve the quality of your soil by adding compost or other organic material. This will loosen clay soils and help sandy soils retain more water. You can also build a raised-bed garden and fill it with commercial garden soil, which is specially formulated for growing plants.

Soil pH: There are some plants that do best in slightly acidic or basic soil. But soil with a neutral pH is going to work for just about everything. Checking and adjusting your soil pH is fairly advanced gardening, so I don't recommend worrying too much about it until you've gotten the basics down. But if your plants aren't thriving, it's a good thing to keep in mind.

Height and spread: It's important not to crowd your plants, so think about their mature size when arranging them in your garden. This will probably leave things looking very sparse at first, but over time you will be glad that you left enough space for them to grow. You also want to consider their mature height. Your garden will look best with the tallest plants in the back, where shorter ones wouldn't be visible.

Annual or perennial: Annuals are plants that only live for one growing season and then die in the winter. **Perennials** will survive the winter and come back year after year, often growing larger with each season. This is important to consider when you plan out your garden, because some of your plants may be there for a very long time!

Hardiness zone: Climate varies greatly across the United States, and plants that can be perennial in one area may not be able to survive the winter in another. The USDA has divided the United States into **plant hardiness zones**, ranging from 1 to 13, to help you determine what you can plant. An equivalent of these zones is also used in many other countries. To find out your hardiness zone, visit planthardiness.ars.usda.gov and input your ZIP code. I will provide the appropriate hardiness zones for the plants below.

Maintaining Your Garden

Mulch: Once your garden is planted, a layer of mulch can help regulate moisture, control weeds, and give the garden a neater look. The layer of mulch should be two to four inches thick (the finer the mulch, the thinner the layer can be). It's best to leave a small ring bare around the base of your plants to let them breathe and discourage pests.

Water: You should water your garden roughly once or twice a week. Water plants close to the base to avoid getting water on the leaves, which can cause fungus and other diseases. Make sure that your soil drains well and that the plants don't remain waterlogged. It's best to let the soil dry out completely between waterings. If you're planting in containers, be sure that they have proper drainage. Overwatering is known as a common issue for potted plants, but I found that my balcony garden needed daily watering in the summer or the plants would begin to droop. It may take some trial and error to learn what is best for yours.

Fertilizer: It's generally a good idea to put down some compost or fertilizer before you plant. Compost is really good for your plants and can drastically improve your soil quality. But it can be a chore to buy the large bags and mix their (often smelly) contents into your soil. When working with good soil to start with, you can also add slow-release fertilizers like Osmocote. These are small beads that you mix into your soil that continuously fertilize your garden.

Once your plants are established, you don't need to fertilize very often—every six weeks is fine for herbs and flowers. Vegetable plants like tomato may benefit from more frequent applications of a fertilizer or plant food.

Deadheading flowers: Deadheading is when you pinch or cut off spent flowers after they bloom, right above the first set of leaves. Most flowering plants will benefit from this—it encourages more blooms and will also prevent self-seeding if that is not desirable. For your annuals, you can save some of the seeds from the flowers you clip to plant the following year.

End-of-season care: When weather gets cold and the growing season has come to an end, there are a number of things you can do to prepare your garden for the winter and the following spring. Pull all the weeds you can find. Remove any dead annuals, saving seeds to plant in the future. Some of your perennials may benefit from pruning or cutting back at this point, while others do better being pruned during the spring or summer—look up the specific needs of each plant. Finally, a layer of mulch can help protect your plants from the coming cold.

Herbs

An herb garden will do double duty, providing ingredients for both cocktails and cooking. There's something extremely satisfying about making a recipe that calls for mint or basil and just walking outside with your shears to harvest some. All of the herbs listed here have both edible leaves and edible flowers. They don't need much water or fertilizer, and most can be harvested throughout the season. They do well in both containers and in gardens. Several are perennials that will come back year after year.

Basil, *Ocimum basilicum*

Delicious and fragrant, basil has many uses in food and drink recipes. The most common variety is sweet basil. Purple basil can also make a lovely garnish. Prune the branches of your basil and pinch off flowering heads to encourage it to spread and produce more leaves, or leave a few if you want flowers for garnishes.

Type: Annual in zones 2–9; perennial in 10–11
Mature size: 1–2 feet tall, 1–2 feet wide
Sun: Full
How to start: Seeds, plants, or cuttings

Chives, *Allium schoenoprasum*

Chive plants produce lovely purple puffball flowers that I love to use as a garnish. The leaves are also a delicious addition to many food recipes, and I used a few to garnish my Bloody Mary (pg. 163). The plants are low-maintenance and very hardy—mine survived the Boston winters even in a container.

Type: Perennial, zones 3–9
Mature size: 12 inches tall, 12 inches wide
Sun: Full
How to start: Seeds, plants, or cuttings

Cilantro, *Coriandrum sativum*

Cilantro isn't an herb you can plant once and then harvest for the rest of the season like basil or chives. It bolts, or blooms, very quickly in hot weather, and once this happens it will produce fewer, more bitter leaves. The best way to have cilantro readily available is to sow new seeds every three weeks and harvest the plants before they bolt. It will also self-seed, so you may find new plants popping up on their own.

Type: Self-seeding annual; plant in spring in zones 3–8 and in fall in zones 9–11
Mature size: 20 inches tall, 12 inches wide
Sun: Full to partial
How to start: Seeds or plants

Dill, *Anethum graveolens*

Dill is another herb with many culinary uses. Like cilantro, it bolts easily, so if you want a constant supply of dill leaves, do several successive plantings. Once it does bolt, its delicate yellow flowers make a unique and beautiful garnish.

Type: Self-seeding annual; may not grow above zone 9
Mature size: 2–4 feet tall, 12 inches wide
Sun: Full
How to start: Seeds, plants, or cuttings

Fennel, *Foeniculum vulgare*

I've always enjoyed the anise-like flavor of fennel, and its fronds and flowers can make dramatic garnishes. Common fennel, which grows tall and does not produce a bulb, comes in green and bronze varieties. If you want a bulb to eat, look for Florence fennel. When it bolts, fennel produces yellow, bouquet-shaped flowers that are similar to those of dill—in fact, they may be able to cross-pollinate, so be careful if you're harvesting seeds.

Type: Usually grown as an annual
Mature size: 3–6 feet tall, 12 inches wide
Sun: Full
How to start: Seeds or plants

Lavender, *Lavandula* spp.

Lavender is one of my favorite plants to grow. This beautiful and fragrant herb produces lovely purple blooms that can be used to flavor food and cocktails. If cared for properly, lavender will grow large and continue to produce blossoms year after year.

Type: Perennial, zones 5–9
Mature size: 1–3 feet tall, 2 feet wide
Sun: Full
How to start: Plants or cuttings

Mint, *Mentha* spp.

Mint is a hardy herb that is easy to grow in a garden or in containers. It spreads readily, to the point where it can take over a garden if not pruned. It's an important ingredient in many classic cocktails. A great choice is Mojito Mint, which comes from Cuba and is the variety of mint first used in the Mojito cocktail.

Type: Perennial, zones 3–9
Mature size: 1–2 feet tall, spreads widely
Sun: Full to partial
How to start: Seeds, plants, or cuttings

Rosemary, *Salvia rosmarinus*

This fragrant evergreen herb is easy to care for and can grow quite large in the garden. A great way to start your own plant is to take a cutting from an established plant and keep it in water until it produces roots. Sprigs of rosemary make festive garnishes for holiday cocktails.

Type: Perennial, zones 7–11
Mature size: 4 feet tall, 4 feet wide
Sun: Full
How to start: Plants or cuttings

Sage, *Salvia officinalis*

Sage is another herb that will get a lot of use in both food and cocktail recipes. There are many varieties, with leaf colors ranging from the typical gray-green to purple or tricolor, which can be particularly nice choices for garnishes.

Type: Perennial, zones 4–10
Mature size: 12–24 inches tall, 12–24 inches wide
Sun: Full
How to start: Seeds, plants, or cuttings

Tarragon, *Artemisia dracunculus*

Tarragon is a lovely herb that doesn't get nearly enough attention. It has an anise-like aroma and savory licorice flavor that make it an incredibly interesting addition to cocktails and food.

Type: Perennial, zones 4–8
Mature size: 2–3 feet tall, 12 inches wide
Sun: Full to partial
How to start: Plants or cuttings

Thyme, *Thymus* spp.

Thyme is a hardy herb that doesn't need much water and stays green through the winter. It can spread widely and makes good ground cover. It produces tiny lavender flowers that can fill in a garnish on a drink or a plate.

Type: Perennial, zones 5–9
Mature size: 6–12 inches tall, spreads widely
Sun: Full to partial
How to start: Plants or cuttings

Fruits and Vegetables

Fruits and vegetables can be a bit more intimidating to plant than herbs or flowers, but they are also extremely rewarding. While you might go out to your herb garden and pluck a few leaves for a garnish, your fruit and vegetable plants can provide a substantial portion of your drinks and meals. Most vegetables and berries are actually quite simple to grow and have a number of uses beyond cocktails. Fruit trees require more of a commitment but will continue to produce for years once they are established.

Apple, *Malus domestica*

Apples come in many different sizes, colors, and flavors, but they don't need a lot of space to thrive. There are dwarf varieties and types that can be grown as an espalier against a fence or trellis for people with limited garden space. Most varieties can't self-pollinate, so be sure to plant at least two in order to produce fruit. They can also be prone to pests, and some degree of pesticide application is usually necessary.

Type: Tree, zones 3-8
Mature size: Varies by variety
Sun: Full
How to start: Plants or cuttings

Bell Pepper, *Capsicum annuum*

Also called sweet peppers, bell peppers lack the capsaicin of spicier varieties. They change color as they ripen, going from green to either red, yellow, or orange, depending on the varietal. A cage or support is recommended to keep the plants from bending as they grow and produce fruit.

Type: Perennial in warm climates, annuals in cold climates, zones 4–11
Mature size: Varies by variety
Sun: Full
How to start: Plants or seeds

Blackberry, *Rubus fruticosus* agg.

Blackberries are extremely simple to grow and can easily take over a garden. The branches, or canes, are biennial—they grow their first year, produce fruit the following year, and then die. There are erect, semi-erect, and trailing varieties that need different amounts of support.

Type: Perennial, zones 4–10
Mature size: 6 feet tall, spreads widely
Sun: Full
How to start: Plants or cuttings

Blueberry, *Vaccinium* sect. *Cyanococcus*

Unlike blackberries and raspberries, blueberries are temperamental and grow slowly. It may be several years before your bush produces a big crop. They prefer acidic soil with a pH of 4.0—5.0, so you can add an acidifier like sulfur to your soil to help your blueberry plants thrive. There are many varieties, some of which can do well in containers. Research the best variety for your zone.

Type: Perennial, zones 3–9
Mature size: 3–6 feet tall
Sun: Full
How to start: Plants or cuttings

Lemons, limes, and oranges, *Citrus* spp.

Although *citrus* grows best in subtropical climates, northern gardeners can grow lemons, limes, oranges, and other citrus trees in containers so they can be moved indoors once the temperature drops too low for them to survive outdoors.

Type: Tree, zones 8-10
Mature size: Varies by variety
Sun: Full
How to start: Plants or cuttings

Cucumber, *Cucumis sativus*

Cucumbers are every gardener's dream. As long as they have plenty of water and sun, they will grow abundantly. Vining varieties can spread widely and do best with a trellis for support, while bush cucumbers are more compact and can be grown in a container. Keep in mind that cucumbers don't like to be transplanted once established.

Type: Annual
Mature size: Varies by variety
Sun: Full
How to start: Plants or seeds

Green onion, *Allium fistulosum*

Green onions are easy to grow and can even be regenerated from the scraps that are cut off of the bottom of the bulb when preparing them. Simply plant the end of the bulb with the roots in soil or place it in water and you'll see the onion quickly start to regrow.

Type: Perennial, zones 5–9
Mature size: 12 inches tall
Sun: Full
How to start: Plants, cuttings, or seeds

Jalapeño pepper, *Capsicum annuum*

Unlike bell peppers, jalapeños have capsaicin, which makes them hot and spicy. They will ultimately turn red if left on the plant long enough, but they're ready to eat once they're a shiny, bright green. Your plant may benefit from a cage for support if it grows large.

Type: Annual
Mature size: 2 to 3 feet tall, 1 foot wide
Sun: Full
How to start: Plants or seeds

Peach, *Prunus persica*

Peach trees vary by zone, so make sure to choose a variety that will thrive where you live. They should be planted while they're dormant and will begin producing fruit within two to four years. If the fruit is abundant, cull down the number of peaches or the branches will break from the weight of the fruit.

Type: Tree, zones 4-9
Mature size: Varies by variety
Sun: Full
How to start: Plants or cuttings

Raspberry, *Rubus* spp.

Like blackberries, raspberries are not hard to grow and can easily take over your garden if you're not careful—this can be especially problematic given their thorny branches. But they will produce abundant, delicious fruit. Be sure to prune back dead canes that have finished fruiting each season to keep your plants neat, productive, and healthy.

Type: Perennial, zones 2–8
Mature size: 3 to 6 feet
Sun: Full
How to start: Plants or cuttings

Rhubarb, *Rheum rhabarbarum*

Rhubarb, though often paired with fruit, berries, and lots of sugar in pie and jams, is actually a vegetable. When the stalks reach eight to ten inches long, they can be harvested by clipping them near the base. You should wait to harvest your rhubarb until the second season after planting, and then only harvest lightly. The following year, you can harvest more heavily. Keep in mind that the leaves are poisonous and should not be consumed.

Type: Perennial, zones 3–8
Mature size: 3 feet tall, 4 feet wide
Sun: Full
How to start: Plants or cuttings

Strawberry, *Fragaria* x *ananassa*

Strawberries are an easy plant to grow, and their fruit is best fresh off the plant. When planting, be sure not to bury the crown of the plant, from which the stems emerge. There are three varieties to choose from: **June-bearing**, which produce large fruit during a couple of weeks early in the season; **everbearing**, which have an early and late crop with a few fruit in between; and **day neutrals,** which produce smaller fruit continuously.

Type: Perennial, zones 2–10
Mature size: 6–8 inches tall, spreads widely
Sun: Full
How to start: Plants, seeds, or cuttings

Sugar Snap Peas, *Pisum sativum* var. *macrocarpon*

Sugar snap peas have an edible pod with a crunchy texture and sweet flavor. They grow on a vine, so provide a trellis or other support and watch as it slowly grows and climbs.

Type: Annual
Mature size: 6–8 feet on a trellis
Sun: Full to partial
How to start: Plants or seeds

Tomato, *Solanum lycopersicum*

Tomatoes are an incredibly easy plant to grow and will produce an abundant amount of fruit. There are many different cultivars, from tiny cherry tomatoes to large, lumpy heirloom varieties. Tomato cages or stakes are a must, as the plants can quickly grow large heavy branches that need support. Keep an eye out for pests and diseases as your plants grow.

Type: Annual
Mature size: Varies by variety
Sun: Full
How to start: Plants or seeds

Flowers

Edible flowers are a stunning addition to cocktails and food. As someone who regularly posts cocktail photos to Instagram, I can tell you that drinks with edible flowers draw far more attention and engagement than any others. It's a bit ironic since the flowers themselves rarely add to the flavor of the cocktail. But a fresh and well-crafted garnish implies that the cocktail is also well-crafted and made with fresh ingredients.

I do often see people on Instagram garnishing their cocktails with any flowers they can find, including toxic or unpalatable varieties. I think this is a very dangerous trend. It's important to do your research and make sure that the flowers you're putting in your drinks are safe. Be aware that some edible flowers can look very similar to nonedible flowers, and make sure that you have identified them correctly before using them in a drink. I don't believe that poisonous plants belong anywhere near your glass. Even if you don't think your guests will eat them, do you really want to have to issue a warning with every drink? Especially when there are so many beautiful edible flowers out there.

Also, remember that, just because something is edible for humans, that does not mean it's safe for your pets to consume (though hopefully you're not letting them anywhere near your cocktails anyway). Begonias are an example of a plant that is fine for us but not safe for our furry friends.

Additionally, just because some parts of a plant are edible doesn't mean that the entire plant can be eaten. Both the leaves and flowers of all of the plants listed below are edible except *dianthus and elderberry*. Dianthus leaves may simply cause an upset stomach, but elderberry leaves and uncooked berries can make you seriously ill.

All of the plants below can be grown in either a garden or a container except lilac, which won't do well in containers, and orchids, which need to be indoors in pots in most hardiness zones.

Aster, *Aster* spp. and *Symphyotrichum* spp.

Asters are small, daisy-like flowers that can be white, pink, purple, or blue. The species native to North America were recently moved into the genus *Symphyotrichum*, while the European and Asian varieties remain in the genus *Aster*. They come in a variety of sizes and heights and are attractive flowers for pollinators.

Type: Perennial, zones 3–8
Mature size: Varies by species
Sun: Full to partial
How to start: Seeds, plants, or cuttings

Bee Balm, *Monarda* spp.

Bee balm is a beautiful, fragrant shrub that produces copious colorful blossoms throughout the summer. It's also known as wild bergamot because the smell of the leaves is reminiscent of the bergamot orange, which is used to flavor Earl Grey tea. The blooms, which can be pink, purple, red, or white, are great for attracting pollinators. It's also a fairly tall shrub, so it's a good choice for the back of your garden.

Type: Perennial, zones 3–9
Mature size: 2–4 feet, spreads widely
Sun: Full
How to start: Seeds, plants, or cuttings

Begonia, *Begonia* spp.

This common houseplant comes in a wide variety of shapes and colors. The flowers, stems, and leaves are all edible—just use caution if you have gout, kidney stones, or rheumatism, as they contain oxalic acid. Because there are so many varieties, check the recommendations of the specific type you choose for information on size and sun preferences.

Type: Annual except in zones 9 and 10
Mature size: 6–12 inches tall, 6–12 inches wide
Sun: Partial sun to shade
How to start: Cuttings or plants

Borage, *Borago officinalis*

Borage is an easy plant to grow from seeds or seedlings. It produces delicate, star-shaped flowers that start out pink and then turn blue. There is also a less-common white variety. The flowers attract a variety of pollinators. Both the flowers and leaves are edible and have a cucumber-like flavor. If it grows large, the branches may become so heavy that it needs a hoop or cage for support. It readily drops large seeds that you can save for the following year.

Type: Self-seeding annual
Mature size: 2–3 feet tall, 12–24 inches wide
Sun: Full to partial
How to start: Seeds or plants

Calendula, *Calendula officinalis*

Calendula is an edible plant that produces cheery, daisy-like flowers in bright orange and yellow. The oil extracted from the flowers is said to have medicinal properties. Like borage, it produces a number of large seeds that are easy to save for future planting.

Type: Self-seeding annual
Mature size: 18–24 inches tall, 1 foot wide
Sun: Full to partial
How to start: Seeds or plants

Chamomile, *Matricaria chamomilla* or *Chamaemelum nobile*

These tiny, daisy-like flowers are most often dried to make tea, but they are also a beautiful and fragrant addition to any garnish. There are two varieties, German and Roman chamomile. Both have a similar appearance, scent, and flavor. German produces more flowers, while Roman makes a better ground cover. Because I'm most interested in the flowers, I have always planted the more common German variety. While it is an annual, its seeds are light and easily carried by the wind. Don't be surprised if you start seeing tiny chamomile plants popping up in unexpected locations the following year!

Type: German is a self-seeding annual; Roman is perennial in zones 3–9
Mature size: German—1–2 feet tall, 1–2 feet across; Roman—3–4 inches tall, spreads widely
Sun: Full to partial
How to start: Seeds or plants

Cornflower, *Centaurea cyanus*

This lovely bloom, in its iconic shade of blue, is a great choice for your garden. It is also commonly called bachelor's button, but so is globe amaranth (see below), which can lead to confusion. Luckily, they are both edible! Most varieties of cornflower are indeed bright blue, but there are also pink, purple, red, and white varieties. Some variants have more compact flowers than others.

Type: Self-seeding annual
Mature size: 1–3 feet, 1–2 feet wide
Sun: Full
How to start: Seeds or plants

Dianthus, *Dianthus* spp.

The genus *Dianthus* encompasses a wide variety of edible flowers, from the flat, five-petaled flowers sometimes called pinks to the showy carnations used in bouquets. It's not uncommon to find several varieties available at your local nursery, so do a bit of research to pick the right one for your garden. Note that the leaves of this plant are not edible.

Type: Varies by species
Mature size: Varies by species
Sun: Full
How to start: Seeds, plants, or cuttings

Elderflower, *Sambucus canadensis*

The American elderberry is a fast-growing shrub that produces tiny, fragrant flowers in June, followed by dark purple elderberries in the late summer. The flowers are edible, but the other parts of the plant are poisonous, so use caution. The berries can be consumed if they have been cooked, which destroys the harmful compounds present in their seeds. It is possible to grow elderberry in a large container, but planting in a garden is ideal.

Type: Shrub, zones 3–10
Size: 6–12 feet tall, 6–12 feet wide
Sun: Full to partial
How to start: Seeds, plants, or cuttings

Fuchsia, *Fuchsia* spp.

I was only recently introduced to this stunning blossom, and it's now one of my favorites. I used it to garnish the Pegu

Club (pg. 125) and the Mock Tai (pg. 141). There are several varieties, including upright, trailing, and climbing fuchsias. The Dollar Princess variety, with hanging pink and purple blooms, is one of the most common and is a popular choice for containers or hanging planters.

Type: Annual; perennial in zones 9 and 10
Mature size: 18 inches tall, 24 inches wide
Sun: Partial shade
How to start: Plants or cuttings

Globe Amaranth, *Gomphrena globosa*

Globe amaranth is a fully edible plant with charming purple, pink, red, or white puffball flowers. It is also commonly called bachelor's button, which can lead to some confusion with cornflower, which shares the nickname. But the two are very different plants. The flowers remain beautiful even when dried, so consider putting a few aside for wintertime garnishes.

Type: Annual
Mature size: 2 feet tall, 1 foot wide
Sun: Full to partial
How to start: Seeds or plants

Hibiscus, *Hibiscus* spp.

There are many types of hibiscus. The tropical varieties you are likely most familiar with can only be grown as annuals (except in zones 9 and 10) and must be brought indoors to survive the winter. But there are also perennial and shrub forms that are hardy down to zone 5. All of the flowers are edible, and will add a bright and tropical touch to your garden.

Type: Varies by species
Mature size: Varies by species
Sun: Varies by species
How to start: Seeds, plants, or cuttings

Lilac, *Syringa vulgaris*

The lilac produces beautiful, fragrant purple blooms each May. The shrubs can grow quite large and live for decades. The bushes should be pruned once they are finished flowering to protect them from pests and prepare them for the next year. They do not do well in containers and should be planted in the garden.

Type: Shrub, zones 3–7
Mature size: 6–16 feet tall, 8–12 feet wide
Sun: Full to partial
How to start: Plants

Marigold, *Tagetes* spp.

The cheery marigold is a common garden plant that is easy to find in nurseries. The most popular species is *Tagetes patula*, the French marigold. They're not known for having a particularly appealing flavor, but they are safe to use as a garnish.

Type: Self-seeding annual
Mature size: 6 inches to 2 feet tall, 6–12 inches wide
Sun: Full
How to start: Seeds, plants, or cuttings

Nasturtium, *Tropaeolum majus*

Nasturtium is a fully edible plant with lily-pad-shaped leaves and bright orange and red flowers. It's a great choice for a new gardener, as it grows very quickly and thrives in poor soil, which will encourage more flowers and fewer leaves. The flowers and leaves have a peppery, vegetal flavor. Aphids are a common pest for nasturtium, so check the stems and undersides of leaves regularly.

Type: Annual
Mature size: 1–2 feet tall, 1–2 feet wide
Sun: Full
How to start: Seeds or plants

Orchid, family *Orchidaceae*

The stunning blooms of orchids are a perfect garnish for Tiki cocktails. While there are varieties that can be grown outdoors in the US, most will do best in pots kept indoors. The most common type, which you may see at your grocery store florist, is the moth orchid (*Phaleonopsis* spp.). To obtain more unusual varieties, like the micro orchids I used in the Daiquiri (pg. 135) and Spicy Avocado Margarita (pg. 154), try a specialized retailer.

Type: Houseplant
Mature size: Varies by species
Sun: Windowsill light
How to start: Plants or cuttings

Violas, *Viola* spp.

The violas include flowers commonly called pansies, violets, and Johnny jump-ups. At the first sign of spring, every nursery and garden store will stock their shelves with trays of these unique and beautiful blooms. They are hardy against the early-season cold, will bloom all summer long, and come in a huge variety of sizes and colors.

Type: Annual
Mature size: 6–10 inches tall, 12 inches wide
Sun: Full to partial
How to start: Seeds, plants, or cuttings

Other Edible Flowers

The section above only covers the flowers I used in this book. There are many more edible flowers that you can grow in your garden! Here are some other varieties to look out for:

Butterfly pea flower, *Clitoria ternatea*—These give indigo gins their color and can be used to infuse other spirits for the same effect.

Buzz buttons, *Acmella oleracea*—These conical yellow blooms cause a tingling sensation in your mouth when eaten.

Chrysanthemum, *Chrysanthemum* spp.—A popular autumn bloom, the chrysanthemum is actually an edible perennial.

Clover, *Trifolium repens*—This species is likely already growing in your area as a weed, and is entirely edible.

Dahlia, *Dahlia* spp.—Both the petals and bulbs of these impressive flowers are edible.

Dandelion, *Taraxacum officinale*—The dandelion is another common edible weed.

Evening primrose, *Oenothera biennis*—All parts of this plant are edible, and the seed oil is thought to have medicinal uses.

Forsythia, *Forsythia x intermedia*—This shrub will be covered in edible yellow flowers in early spring.

Geranium, *Pelargonium* spp.—The flowers and leaves of this very popular garden plant are edible.

Honeysuckle, *Lonicera japonica*—Only the flowers of this fragrant climbing vine are edible.

Hosta, *Hosta* spp.—This popular garden plant, usually chosen for its attractive leaves, also produces edible flowers.

Kousa dogwood, *Cornus kousa*—The flowers and fruit of the Kousa dogwood are edible, but be wary, as most dogwoods are poisonous.

Jasmine, *Jasminum sambac*—Be careful to look for Arabian jasmine, as other varieties are not edible.

Ornamental onion, *Allium* spp.—The alliums produce dramatic, tall stalks with globular clusters of flowers.

Peony, *Paeonia* spp.—These beautiful blooms are also a great choice for bouquets and cut flowers.

Rose, *Rosa* spp.—All roses are edible.

Sunflower, *Helianthus annuus*—In addition to the seeds, the petals of a sunflower can be eaten.

The Recipes

Cognac

Brandy Crusta

Classic

2 oz. Cognac

½ oz. orange Curaçao

½ oz. maraschino liqueur

½ oz. lemon juice

2 dashes Angostura bitters

Pro tip: Peel your lemon before you juice it. With a sharp knife or Y-peeler, begin at the top of your lemon, removing a section of peel about an inch wide, slowly moving down and around the lemon, until you have removed the entire peel. It should naturally coil in the top of your glass.

With its sugar crust and dramatic lemon peel garnish, the Brandy Crusta exudes the elegance of old New Orleans, where it was invented in 1852.

First, prepare your glass. Rub a slice of lemon along the side of a champagne flute or small wine glass and roll it in sugar to coat. Combine all ingredients for the cocktail in a shaker with ice and shake until chilled. Strain into the prepared glass. Garnish with a whole lemon peel curled around the inside of the glass.

Garnish: *sugared rim, whole lemon peel, viola*

Rosemary Pear Crusta

Original

Rosemary is one of my favorite garnishes because it's so fragrant. Just a small sprig—or, in this case, a hint of rosemary oil in a sugared rim—lends a huge amount of aroma and flavor to a cocktail.

Prepare your glass by rubbing a lemon slice along the outside and rolling it in rosemary sugar. Combine all cocktail ingredients in a shaker with ice and shake until chilled. Strain into the prepared glass.

Rosemary sugar: In a small bowl, combine sugar and rosemary leaves and stir gently to spread the oils from the leaves to the sugar. Pour into a small dish to rim your glass.

2 oz. Cognac

¼ oz. maraschino liqueur

1 oz. pear juice

½ oz. lemon juice

2 dashes orange bitters

Rosemary Sugar:

¼ cup granulated sugar

1 tsp. small, fresh rosemary leaves

Garnish: *rosemary sugar rim, pear slice, rosemary*

Sidecar

Classic

2 oz. Cognac
¾ oz. triple sec
¾ oz. lemon juice
¼ oz. simple syrup

Though its exact origins are uncertain, the popular story is that the Sidecar was created by an American army captain who frequented the famous Harry's New York Bar in Paris and would come and go in a motorcycle sidecar.

Prepare a coupe glass by running a slice of lemon along the rim and dipping it in sugar to coat. Combine all the cocktail ingredients in a shaker with ice and shake until chilled. Strain into the prepared glass.

Garnish: *sugared rim, cornflower petals*

Garden State

Mocktail

Shrubs, also called drinking vinegars, are mixtures of fruit, sugar, and vinegar that can be enjoyed in cocktails or mocktails. The process of making them dates back to colonial times, when vinegar was used as a preservative for fruits and berries.

Prepare a Collins glass by running a slice of lemon along the rim and rolling the glass in sugar. Fill the glass with ice. Combine simple syrup and tarragon in the bottom of a shaking tin and muddle gently. Add lemon juice and satsuma shrub. Fill shaker with ice and shake until chilled. Fine strain into the prepared glass and top with club soda.

Satsuma Shrub: With a grater, zest the satsumas, collecting the grated peel in a large bowl. Add sugar and stir, pressing the peel into the sugar to impart the oils. Chop the satsumas into chunks and add them to the bowl. Stir, pressing the satsuma pieces with your spoon, until all of the sugar is dampened by the juices. Cover the mixture with a kitchen towel and let sit for two days, stirring at least once a day. After two days, the sugar should be liquified into a syrup. Fine strain the syrup into a measuring cup.

1½ oz. chilled rooibos tea

1 oz. satsuma shrub

¾ oz. lemon juice

½ oz. simple syrup

1 sprig fresh tarragon

3 oz. club soda

Satsuma Shrub:

½ lb. satsumas (about 2)—can substitute mandarin oranges, blood oranges, or anything similar

1 cup sugar

½ cup apple cider vinegar

Pro tip: It's easy to use this basic recipe (½ lb. fruit, 1 cup sugar, ½ cup vinegar) to make a shrub out of anything! Mix your shrub with club soda and you have an easy and impressive drink. You can also add your favorite spirit for a tangy, delicious cocktail.

You should have about 1 cup. Add the apple cider vinegar, a little at a time, tasting as you go for a good balance of tangy and sweet. I like to add half as much vinegar as there is syrup.

Garnish: *tarragon, aster*

The Recipes

Gin

Bee's Knees

Classic

2 oz. gin
¾ oz. honey syrup
¾ oz. lemon juice

Honey Syrup:
½ cup honey
½ cup water

This drink dates back to the Prohibition era, when calling something the "bee's knees" meant it was the best. It's also a reference to the honey syrup in the drink. In terms of effort vs. impact, honey syrup is hard to beat—try using it instead of simple syrup to dress up other recipes!

Combine all ingredients in a shaker with ice and shake until chilled. Strain into a coupe glass.

Honey syrup: Combine in a cup or bowl and stir until the honey dissolves.

Garnish: *honeycomb, lilac, bee pollen*

Daisy Chain

Mocktail

Tea is a perfect base for nonalcoholic drinks. This super-refreshing mocktail combines the flavors of chamomile, apple, and honey.

Combine all ingredients in a shaker with ice and shake until chilled. Strain into a rocks glass—or honey jar!—filled with ice.

To make your own apple juice: Cut apples into slices (no need to peel or core them) and place them in a large saucepan. Add enough water to just cover the apple slices. Bring the mixture to a boil and let it simmer for about twenty minutes, until the apples are very soft. Mash the apples with a potato masher or the back of your spoon. Then strain the mixture through a fine mesh strainer or cheesecloth. Add sugar if desired.

2 oz. chamomile tea, cooled

1 oz. apple juice

¾ lemon juice

¾ honey syrup

Garnish: *Chamomile flowers*

Clover Club

Classic

1½ oz. gin

½ oz. dry vermouth

½ oz. lemon juice

½ oz. raspberry syrup

¼ oz. egg white

Raspberry Syrup:

¼ cup raspberries

½ cup sugar

¼ cup water

Pro tip: *Egg white is a very common addition to classic sour cocktails. It doesn't add much flavor; rather, it gives the cocktail a creamy, silky texture and a lovely layer of foam on top. There's always a small risk when consuming raw eggs, so make sure yours are fresh. I like to separate my egg white into a bowl and lightly whisk it so that it's easier to measure out ¼ ounce.*

Named for the men's club in Philadelphia where it originated, the Clover Club is a raspberry gin sour that's as delicious as it is beautiful. This recipe comes from Julie Reiner, who is such a fan that she named her now-famous Brooklyn bar after the drink.

Do a reverse dry shake: combine all ingredients in a shaker with ice and shake well, for at least twenty seconds. Strain the drink, dump the ice, and return the cocktail to the shaker to shake again for at least thirty more seconds. Strain into a coupe.

Raspberry Syrup: Muddle raspberries in a bowl. Add the sugar and stir or muddle to mix it in well. The mixture should become bright red and juicy. Let it macerate for twenty to thirty minutes. Then add the water, stir well, and fine strain.

Garnish: *raspberries*

Carrie Nation

Mocktail

Aquafaba—the liquid from canned chickpeas— is a vegan egg white substitute. It's a great way to make pregnancy-friendly mocktails that still look like impressive cocktails.

Do a reverse dry shake: combine all ingredients in a shaker with ice and shake well, for at least twenty seconds. Strain the drink, dump the ice, and return the cocktail to the shaker to shake again for at least thirty more seconds. Open pour (without a strainer) into a large coupe.

Mixed Berry Syrup: Combine the berries in a bowl and muddle well. Add the sugar, stirring or muddling until it is mixed in well. Let the mixture sit for thirty minutes, then stir in the water and fine strain.

3 oz. cold fruity tea (I used Tazo Passion)

1 oz. lemon juice

1 oz. mixed berry syrup

1 oz. aquafaba

Mixed Berry Syrup:

1 cup mixed berries (raspberries, blackberries, blueberries, strawberries)

2 cups sugar

1 cup water

Garnish: *strawberries, blackberries, raspberries, blueberries*

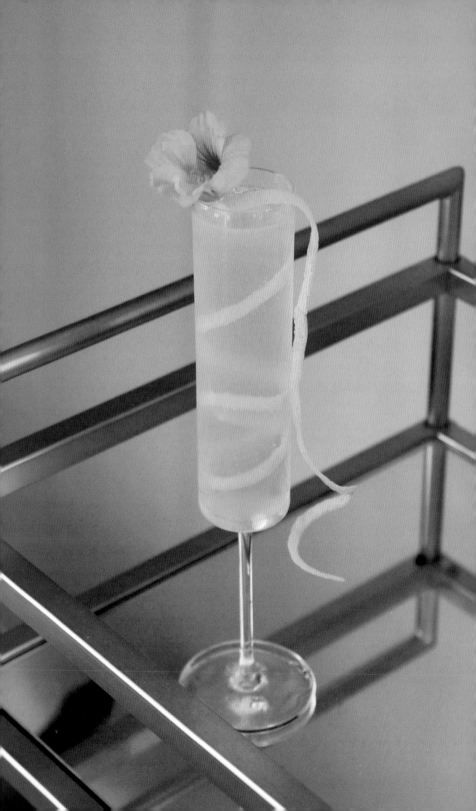

French 75

Classic

2 oz. gin or Cognac

½ oz. lemon juice

1 tsp. simple syrup

Sparkling wine, to top

There are few cocktails as elegant as the French 75. Though it's most commonly made with gin, there is also a version made with Cognac. Just be careful—there's a reason this drink is named after heavy artillery!

Combine gin or Cognac, lemon juice, and simple syrup in a shaker with ice and shake until chilled. Strain into a champagne flute and top with sparkling wine.

Mocktail option: Nonalcoholic sparkling wines can be an elegant substitute for the real thing. To make an alcohol-free French 75, eliminate the gin and replace the sparkling wine with a nonalcoholic version. You can also replace the simple syrup with juniper syrup (pg. 114) to keep some of the flavor of the gin.

Garnish: *lemon twist, nasturtium*

Lavande 75

Original

The way a cocktail looks can affect our impression of how it tastes. Using an indigo gin like Empress 1908 for this French 75 emphasizes the inclusion of lavender, which isn't enough to turn the cocktail purple on its own.

Combine gin, lemon juice, and lavender syrup in a shaker with ice and shake until chilled. Strain into a champagne flute and top with sparkling wine.

Lavender Syrup: Combine all ingredients in a saucepan and bring to a simmer. Cook, stirring, until sugar is dissolved. Cover and remove from heat. Let steep until cooled. Fine strain.

Mocktail option: As with the French 75, the Lavande 75 can be made nonalcoholic by eliminating the gin and using an alcohol-free sparkling wine.

2 oz. gin

½ oz. lemon juice

¼ oz. lavender syrup

Sparkling wine, to top

Lavender Syrup:

½ cup sugar

½ cup water

1 tbsp lavender flowers

Garnish: *lavender flower*

Gimlet

Classic

2 oz. gin

¾ oz. lime juice

¾ oz. simple syrup

A Gimlet is a simple but delicious cocktail that should be in everyone's repertoire. You can also make it with vodka.

Combine ingredients in a shaker with ice and shake until chilled. Strain into a coupe glass.

Garnish: *lime slice, kinome leaf, micro princess flower, purple basil blossom*

Snap Judgement
Original

Peas may not seem like a traditional cocktail ingredient, but when paired with fresh mint, they make a classic Gimlet even more refreshing.

Combine peas, mint, and simple syrup in the bottom of a shaking tin and muddle well, crushing the peas. Add lime juice and gin. Fill shaker with ice and shake until chilled. Fine strain into a coupe.

2 oz. gin

¾ oz. lime juice

¾ oz. simple syrup

5 sugar snap peas

5 mint leaves

Garnish: *sugar snap pea, viola, mint leaf*

Gin & Tonic

Classic

2 oz. gin

3½ oz. tonic water

Citrus slices, herbs, and/or spices for garnish

Pro tip: *The garnish plays a large part in the flavor of a Gin & Tonic. Some of my favorite additions include citrus such as lemon, lime, or grapefruit; cucumber; jalapeño; fresh herbs such as mint, basil, or rosemary; and spices such as peppercorns, star anise, or cloves.*

In the nineteenth century, during the British colonization of India, British soldiers took quinine as a prophylaxis against malaria. Because it was quite bitter, they mixed it with sugar, soda water, lime, and their ration of gin to make it more palatable. And thus the Gin & Tonic was born.

Pour gin into a rocks glass. Add ice and garnishes. Top with tonic water and enjoy.

Garnish: cucumber, mint, rosemary, jalapeño, viola, lime

Juniper & Tonic

Mocktail

I developed this recipe when I was pregnant with my son and started craving gin. Tonic water and lime is a refreshing drink, but it's missing the juniper flavor that dominates most classic gins. Luckily, dried juniper berries aren't too hard to find—my local big-chain supermarket had them in the spice section— and they bring this mocktail as close to a G&T as you can come without alcohol.

Combine juniper syrup and lime juice in a rocks glass. Fill glass with ice and top with tonic water. Stir very briefly.

Juniper Syrup: Combine sugar, water, and juniper berries in a small saucepan and muddle the juniper berries a bit to release their flavor. Add the remaining ingredients and bring to a simmer. Reduce heat to low, cover, and simmer for fifteen minutes. Let cool. Transfer to a jar or other container and refrigerate a few hours or overnight, then strain out the spices.

1 oz. juniper syrup

½ oz. lime juice

5 oz. tonic water

Juniper Syrup:

½ cup sugar

½ cup water

2 tbsp. dried juniper berries

1 strip of orange peel

1 cardamom pod

1 small bay leaf

Garnish: *chamomile flowers, chive blossoms, juniper berries, lime*

Martini

Classic

2½ oz. gin
¾ oz. dry vermouth

A Martini is simply gin (or vodka) and dry vermouth, garnished with olives or a lemon twist (or both). The ratio of these two ingredients is a matter of personal taste. When you hear someone order a Martini "dry," this means they don't want very much vermouth, where as a "wet" Martini contains more. This recipe uses a pretty classic ratio.

Combine gin and vermouth in a mixing glass. Add ice and stir. Strain into a chilled martini glass or coupe.

Garnish: *olives, lemon twist*

Dirty Gibson
Original

A Gibson is a Martini that is garnished with a cocktail onion instead of an olive or lemon twist, and a "dirty" Gibson or Martini includes a bit of brine from the onion or olive jar in the mix as well. For this Dirty Gibson, I pickled my own veggies in a homemade lemon-herb brine, which then goes into the drink as well.

Combine ingredients in a mixing glass with ice and stir until chilled. Strain into a chilled cocktail glass and garnish with an onion or scallion pickled in lemon-herb brine.

Lemon-Herb Brine: Combine vinegar, sugar, water, and salt in a saucepan and bring to a simmer. Add remaining ingredients and stir. Remove from heat and cover, letting cool completely.

Pour the cooled brine into a jar and add onions and anything else you'd like to pickle. Let sit at room temperature for two hours, then transfer to the refrigerator overnight.

2½ oz. gin

½ oz. dry vermouth

1 tsp. lemon-herb brine

Lemon-Herb Brine:

2 cups champagne vinegar

½ cup sugar

¼ cup water

½ tbsp. salt

½ tbsp. mustard seed

1 tsp. coriander seed

1 tsp. peppercorns

2 cloves garlic, sliced

1 lemon, sliced

Several sprigs fresh dill, fennel, and thyme, half of them chopped (about ½ cup of herbs total)

Garnish: *pickled onion, pickled scallion*

Negroni

Classic

1 oz. gin

1 oz. sweet vermouth

1 oz. Campari

The Negroni was famously created by an Italian count named Camillo Negroni, who marched into Florence's Caffe Casoni in 1919 and asked for an Americano (Campari, sweet vermouth, and soda water) with gin instead of soda. The simple 1:1:1 recipe is easy to remember.

Combine ingredients in a mixing glass with ice and stir. Strain into a rocks glass over one large cube. Express an orange peel over the drink.

Garnish: *dried mandarin, viola*

Strawberry Blonde

Original

Infusions are a great way to add flavor to a cocktail. This Negroni is made with strawberry-infused Campari and blanc vermouth to give it a lighter, fruitier flavor.

Combine ingredients in a mixing glass with ice and stir until chilled. Strain into a rocks glass over one large ice cube.

Strawberry-Infused Campari: Combine in a jar and let sit overnight, shaking occasionally. Fine strain before using.

1 oz. gin

1 oz. strawberry-infused Campari

1 oz. blanc vermouth (can substitute sweet vermouth)

Strawberry-Infused Campari:

½ cup sliced strawberries (3 large strawberries)

¾ cup Campari

Pegu Club

Classic

2 oz. gin

1 oz. orange Curaçao

¾ oz. lime juice

1 dash Angostura bitters

1 dash orange bitters

The Pegu Club hails from Myanmar, where it was created at a British gentleman's club in Rangoon (now Yangon) in the 1920s.

Combine all ingredients in a shaker with ice and shake until chilled. Strain into a coupe glass.

Garnish: *fuchsia, lime twist*

Irrawaddy Float

Original

This playful take on the Pegu Club replaces the orange liqueur with a scoop of orange sherbet for a creamy, tangy summer cocktail.

Scoop orange sherbet into a chilled coupe glass. Combine gin, lime juice, and simple syrup in a shaker with ice and shake until chilled. Strain into the coupe over the sherbet. Add club soda and dash bitters on top. Serve with a spoon.

2 oz. gin

¾ oz. lime juice

¾ oz. simple syrup

1–2 oz. club soda

1 scoop orange sherbet

1 dash Angostura bitters

Mocktail option: The gin can be eliminated from the Irrawaddy Float to make it essentially nonalcoholic. You can use juniper syrup (pg. 114) instead of simple syrup to make it taste more like gin. If you are avoiding alcohol altogether, you should also remove the Angostura bitters.

Tom Collins

Classic

1 ½ oz. gin
¾ oz. lemon juice
¾ oz. simple syrup
Club soda

Pro tip: *Flowers will show up best in clear ice. See the Ice section of Bar Basics for tips on how to achieve this.*

This is a simple and very refreshing classic. It's a great choice for anyone who wants the ease and fizz of a Gin & Tonic without the bitter flavor.

Combine gin, lemon juice, and simple syrup in a shaker with ice and shake until chilled. Strain into a Collins glass filled with ice. Top with club soda.

Mocktail option: Eliminate the gin and replace the simple syrup with juniper syrup (pg. 114).

Garnish: *flowers frozen in ice cubes*

Cucumber Collins

Original

*This is one of my favorite summer cocktails.
It's incredibly refreshing and utterly delicious.*

Combine gin, elderflower liqueur, lime juice,
simple syrup, and cucumber water in a shaker
with ice and shake until chilled. Strain into
a Collins glass filled with ice and top with
club soda.

For cucumber juice: Grate cucumber into a
fine strainer and press to release the water.
Alternatively, muddle and then strain your
many failed attempts at creating the perfect
cucumber ribbon.

Mocktail option: Eliminate the gin and
elderflower liqueur. Replace the simple syrup
with juniper syrup (pg. 114) and the club soda
with elderflower soda.

1½ oz. gin

¼ oz. elderflower liqueur

¾ oz. lime juice

½ oz. simple syrup

½ oz. cucumber juice

2 oz. club soda

Garnish: *cucumber ribbon, elderflower*

The Recipes

Rum

Daiquiri

Classic

2 oz. white rum
¾ oz. lime juice
¾ oz. simple syrup

The Daiquiri is one of the bartending industry's most beloved cocktails, but it is rarely made correctly by those outside of it. A true Daiquiri is the essence of simplicity: rum, lime, and sugar. If you try only one recipe in this book, make a proper Daiquiri! It may become your new favorite.

Combine all ingredients in a shaker with ice and shake until chilled. Strain into a coupe glass.

Garnish: *lime wheel, micro orchid*

Sergeant Pepper

Original

Bell pepper adds a bright, vegetal flavor to a classic Daiquiri that blends so well with rum and lime that it's hard to recognize in the cocktail.

Combine bell pepper and honey syrup in the bottom of a shaker and muddle. Add lime juice and rum. Fill the shaker with ice and shake until chilled. Strain into a coupe.

2 oz. white rum

¾ oz. honey syrup

¾ oz. lime juice

2 slices green bell pepper

Garnish: *bell pepper, bachelor button*

Mai Tai

Classic

The Mai Tai is an iconic Tiki cocktail that was invented by Victor "Trader Vic" Bergeron. The story goes that he served it to a Tahitian friend who tried it and exclaimed "Maita'i roa ai!" or, "Very good!" For an authentic garnish, use a spent lime shell and a sprig of mint to represent an island and a palm tree.

Combine all ingredients in a shaker with crushed ice and shake briefly. Pour the entire contents of the shaker into a rocks glass. Heap more crushed ice on top and serve with a straw.

Orgeat: Preheat your oven to 400°F. If your almonds are raw, blanch them by placing them in boiling water for one minute. After a minute, strain them out and run cold water over them.

Spread almonds on a baking sheet and toast them in the oven for about four minutes, shaking the pan halfway through, until they are fragrant but not browned. Let them cool, then add them to a food processor. Blend until they are finely ground.

2 oz. aged rum

½ oz. orange Curaçao

1 oz. lime juice

¼ oz. orgeat

¼ oz. simple syrup

Orgeat:

2 cups blanched almonds

1½ cups sugar

1¼ cups water

1 oz. high-proof rum or other spirit

Pro tip: *Trader Vic used 1 oz. of rum from Martinique and 1 oz. of rum from Jamaica for his Mai Tai. One great bottle to look for is Denizen Merchant's Reserve, which is a mixture of rums from these two islands specially blended for Trader Vic's recipes. Otherwise, try 1 oz. each of Rhum Clement and Appleton Estate to make your Mai Tai completely authentic.*

Combine water and sugar in a saucepan and bring to a simmer, stirring until the sugar is dissolved. Add the almonds and continue to simmer, stirring occasionally, until the mixture is about to boil. Remove from the heat and cover. Let sit for three to twelve hours.

Strain the mixture through cheesecloth. Add the rum and stir.

Garnish: *spent lime shell, mint sprig, orchid*

Mock Tai

Mocktail

4 oz. orange juice

1½ oz. lime juice

¾ oz. orgeat

¾ oz. simple syrup

Crushed pineapple ice

Crushed Pineapple Ice:

6 oz. pineapple juice

6 oz. water

Pineapple juice isn't an ingredient in the classic Mai Tai, but it can be a delicious addition. This nonalcoholic version is made with crushed pineapple ice cubes that slowly change the flavor of the drink as they melt.

Combine orange juice, lime juice, orgeat, and simple syrup in a shaker with ice and shake until chilled. Strain into a large rocks glass or snifter filled with the crushed pineapple ice. Serve with a straw.

Crushed pinapple ice: In a measuring cup with a spout, mix the pineapple juice and water. Fill an ice cube tray with the mixture and let it freeze. When the cubes are solid, crush them in a Lewis bag or plastic bag.

Garnish: *pineapple leaves, pineapple slice, fuchsia*

Mojito

Classic

It's hard to find a cocktail more refreshing than this Cuban classic.

Place mint leaves and simple syrup at the bottom of a Collins glass and muddle gently. Add lime juice and rum. Fill glass with ice and top with club soda. Stir briefly. Serve with a straw.

Mocktail option: A Mojito is still refreshing and delicious with the rum removed. Increase the lime juice and simple syrup to 1 oz. and add club soda to taste.

2 oz. white rum
¾ oz. lime juice
¾ oz. simple syrup
8–10 mint leaves
3 oz. club soda

Garnish: *mint, mint flower*

Frojito
Original

2 oz. white rum

1 oz. lime juice

¾ oz. mint syrup

About a cup of ice

Mint Syrup:

1 cup sugar

1 cup water

1 cup fresh mint leaves

A byproduct of the Frosé craze, the Frojito started showing up on summer patio cocktail menus a few years ago. It's an easy and impressive drink that will become your new favorite summer treat.

Combine all ingredients in a blender and blend until smooth. Pour into a glass and serve with a straw.

Mint Syrup: Combine sugar and water in a saucepan and bring to a simmer, stirring occasionally, until the sugar is dissolved. Add mint and stir. Remove from the heat and let steep, covered, for fifteen minutes. Strain out the mint and let the syrup cool completely before using.

Mocktail option: Leave out the rum and use 1½ oz. each of lime juice and mint syrup (or to taste).

Garnish: *mint*

Old Cuban

Classic

The Old Cuban is actually not old at all—it was invented in the early 2000s by Audrey Saunders of Pegu Club in New York City. It's considered by many to be a "new classic," and it's one of my all-time favorite cocktails.

Combine mint and simple syrup in the bottom of a shaker tin and muddle gently. Add rum, lime juice, and bitters. Fill the shaker with ice and shake until chilled. Fine strain into a coupe glass and top with sparkling wine.

1½ oz. aged rum

¾ oz. simple syrup

¾ oz. lime juice

2 dashes Angostura bitters

6 mint leaves

Sparkling wine, to top

Garnish: *mint, dianthus*

Nuevo Cubano

Mocktail

9 mint leaves

1 oz. simple syrup

1½ oz. lime juice

Nonalcoholic sparkling wine, to top

Pro tip: *Adjust the amount of simple syrup depending on the sweetness of your nonalcoholic sparkling wine. My current favorite is St. Regis Nosecco, which is nicely tart and reminds me more of the real thing than any others I've tried.*

It is miraculously easy to turn the Old Cuban into an impressive mocktail. When I was pregnant with my son, this was my favorite thing to drink to make me feel like I was having a fancy cocktail.

Combine mint leaves, lime juice, and simple syrup in a shaker and muddle gently. Add ice and shake until chilled. Fine strain into a coupe glass and top with nonalcoholic sparkling wine.

Garnish: *begonia*

The Recipes

Tequila

Margarita

Classic

2 oz. blanco
tequila

1 oz. triple sec

1 oz. lime juice

¼ oz. agave nectar

A well-crafted Margarita can transcend anything you've ever poured out of a pitcher in a Tex-Mex restaurant. When I tried the recipe below, the Margarita became one of my favorite cocktails again.

If desired, rim a rocks glass by rubbing a slice of lime along the edge and dipping it in salt. Fill the glass with ice. Combine all ingredients in a shaker with ice and strain into the prepared glass.

Garnish: *lime, hibiscus, purple basil flower*

Spicy Avocado Margarita

Original

I absolutely love cilantro, and it is delicious muddled into a Margarita (pg. 153), Daiquiri (pg. 135), or Gimlet (pg. 109). This drink pairs it with avocado and jalapeño for a creamy, subtly spicy cocktail.

Combine avocado, cilantro, lime juice, and agave nectar in a blender and blend until smooth. The mixture will be very thick. Add tequila and triple sec and blend very briefly to combine. Pour into a shaker and shake with ice until chilled. Strain into a rocks glass filled with ice.

Jalapeño-Infused Blanco Tequila: Combine the tequila and jalapeño in a jar. Seal and shake lightly. Let sit twenty-four hours, shaking occasionally, and then strain out the jalapeño. For more spice, add more jalapeño and/or let it sit longer. For less, use less jalapeño or remove the seeds before you do the infusion. The infused tequila should last indefinitely.

2 oz. jalapeño-infused blanco tequila

¾ oz. lime juice

½ oz. triple sec

½ oz. agave nectar

1 slice avocado

½ tbsp cilantro leaves

Jalapeño-Infused Blanco Tequila:

1 cup blanco tequila

½ a jalapeño, sliced

Garnish: *micro orchids*

Paloma

Classic

2 oz. blanco tequila

2 oz. grapefruit juice

½ oz. lime juice

½ oz. simple syrup

2 oz. club soda

The Paloma is a perfect summer cocktail that deserves a place beside the Margarita as a crowd-pleasing tequila classic.

Prepare a Collins glass or rocks glass by rubbing a wedge of lime along the outside and rolling the glass in salt. Fill the glass with ice. Combine tequila, grapefruit juice, lime juice, and simple syrup in a shaker with ice and shake until chilled. Strain into the prepared glass and top with club soda.

Garnish: *bee balm, grapefruit*

Flower Crown

Mocktail

Grapefruit, rosemary, and cinnamon are an unexpectedly wonderful flavor combination. This syrup would be just as delicious in a traditional Paloma, but here it makes a grapefruit mocktail particularly special.

In a rocks glass, combine grapefruit juice, lime juice, and cinnamon-rosemary syrup. Stir briefly. Add one large ice cube and top with club soda. Garnish with a grapefruit slice and a sprig of rosemary.

Cinnamon-Rosemary Syrup: Combine sugar and water in a small saucepan and bring to a simmer, stirring, until the sugar is dissolved. Add cinnamon and rosemary and simmer for another minute or two. Remove from the heat, cover, and let steep for thirty minutes before straining.

1½ oz. grapefruit juice

½ oz. oz. lime juice

¾ cinnamon-rosemary syrup

2 oz. club soda

Cinnamon-Rosemary Syrup:

½ cup sugar

½ cup water

3 cinnamon sticks

2 sprigs rosemary

Garnish: *grapefruit, rosemary, borage, marigold, dill flower*

The Recipes

Vodka

Bloody Mary

Classic

2 oz. vodka

6 oz. tomato juice

2 dashes Tabasco sauce (or more to taste)

2 dashes Worcestershire sauce

¼ tsp. horseradish

1 pinch celery salt or ground celery seed

1 pinch Tony Chachere's Creole Seasoning

1 pinch ground black pepper

1 pinch salt

Generous squeeze of lemon juice

A Bloody Mary is one drink where it's acceptable to go absolutely over-the-top with your garnish, so don't hold back. I like to pair the usual celery and olives with cheese and fresh herbs.

Combine all ingredients in a tall glass and stir well. Add ice.

Garnish: parsley, dill, chives, calendula, tomato, cheese, pickle, olive, celery

Thai Bloody Mary

Original

Pungent fish sauce gives this Thai-inspired Bloody Mary a huge umami kick.

2 oz. vodka

6 oz. tomato juice

½ oz. lime juice

½ oz. fish sauce

¼ cup fresh cilantro leaves

½ inch section of serrano pepper, sliced

Combine tomato juice, lime juice, fish sauce, cilantro, and serrano pepper in a blender or food processer and blend until the cilantro and pepper are pureed into tiny pieces. The mixture may foam a bit; let it settle and then stir gently to recombine the ingredients.

To prepare the glass, brush the outside of a rocks glass with lime juice and roll in a Cajun seasoning such as Tony Chachere's. Fill the glass with ice and add the vodka. Pour in the tomato mixture and stir.

To make your own tomato juice: Cut fresh tomatoes into quarters and place them in a saucepan over medium heat. If desired, you can add onions or celery as well. Bring to a simmer and cook, mashing the tomatoes as you go, until they are soft and breaking down. Run the entire mixture through a sieve or strainer to extract the juice. Add salt, sugar, spices, and/ or lemon juice if desired.

Garnish: *cilantro, pickled bean sprouts, shrimp, Cajun seasoning*

Cosmopolitan

Classic

1½ oz. vodka

¾ oz. triple sec

1 oz. cranberry juice

¾ oz. lime juice

Pro tip: *The original Cosmo recipe called for an orange vodka such as Absolut Citron—try this for an extra burst of citrus flavor.*

This Carrie Bradshaw favorite was one of the few craft cocktail recipes to come out of the 1990s and stick.

Combine all ingredients in a shaker with ice and shake until chilled. Strain into a cocktail glass or coupe.

Garnish: *dianthus*

In Vogue
Mocktail

Infusing a cranberry syrup with orange peel lends a depth of flavor to this fizzy mocktail that is reminiscent of orange liqueur.

Add cranberry-orange syrup and lime juice to the bottom of a Collins glass. Fill with ice and stir briefly. Top with club soda.

Cranberry-Orange Syrup: Combine sugar and cranberry juice in a small saucepan. Bring to a simmer over medium-low heat, stirring until the sugar is dissolved. Add the orange peel and simmer for three more minutes, stirring occasionally and pressing on the peels with the spoon to release their oils. Cover and remove from heat. Let steep thirty minutes. Strain and let cool before using.

1 oz. cranberry-orange syrup

1 oz. lime juice

4 oz. club soda

Cranberry-Orange Syrup:

½ cup sugar

½ cup cranberry juice

Peel of 1 mandarin orange, torn into pieces

Garnish: *orchid, basil leaf*

Moscow Mule

Classic

1½ oz. vodka
¾ oz. lime juice
¾ oz. simple syrup
3 oz. ginger beer

The Moscow Mule was the product of a fortuitous meeting at a Hollywood restaurant called the Cock'n Bull in 1941. The restaurant's owner was trying to sell his homemade ginger beer, and he met a couple of liquor importers trying to convince Americans to drink their vodka. They came up with a cocktail that would eventually put both products on the map. They served it in a signature copper mug to make it recognizable. Even though it originated as an advertising gimmick, the Moscow Mule has become a beloved classic.

Combine the vodka, lime juice, and simple syrup in a shaker with ice and shake until chilled. Strain into a copper mug filled with ice and top with the ginger beer. Garnish with a lime wheel.

Garnish: *lime, begonia, micro princess flower*

Mocktail Mule

Mocktail

The Moscow Mule is a perfect template
for mocktails since most of its flavor comes
from its nonalcoholic ingredients. To make a
nonalcoholic version more exciting than the
original, I love to add fresh herbs like basil and
switch out the simple syrup for honey or agave.

Combine basil, lime juice, and agave nectar
in the bottom of a mixing glass or tin. Gently
muddle the basil leaves. Add ice and stir
briefly. Fine strain into a copper mug filled
with ice. Top with ginger beer. Garnish with
fresh basil.

5–6 basil leaves

¾ oz. lime juice

½ oz. agave nectar

4 oz. ginger beer

Garnish: *purple basil flower*

The Recipes

Whiskey

Manhattan

Classic

2 oz rye or bourbon whiskey

1 oz. sweet vermouth

1 dash Angostura bitters

Pro tip: *Toss those neon-red maraschino cherries and look for a jar of brandied cherries like the ones made by Luxardo. They're carried in many liquor stores and grocery stores. It's a serious upgrade, and eating them might just be the best part of your cocktail.*

The Manhattan is an iconic cocktail that everyone should be able to make. Vermouth plays a big role in this drink, so be sure that yours is fresh and has been stored in the refrigerator.

Combine all ingredients in a mixing glass with ice and stir until chilled. Strain into a coupe.

Garnish: *orange peel, brandied cherries*

Upstate
Original

Caramelizing or grilling fruits before adding them to your drink can give bright, tart ingredients like rhubarb a deeper, sweeter flavor and a hint of smoke.

Place rhubarb in the bottom of a mixing glass and muddle well, until it is smashed and has released its juices. Add remaining ingredients and fill the mixing glass with ice. Stir until chilled, then fine strain into a coupe glass.

To make caramelized rhubarb: Take a 3-inch piece of rhubarb and split it in half lengthwise. Dredge it in sugar to coat. Ideally, place it on a hot grill—this will infuse it with extra smoky flavor. Alternatively, a crème brûlée torch will do the job. Cook the rhubarb until it is soft and charred and the sugar has become caramelized. Let cool before adding to the cocktail.

2 oz. bourbon or rye whiskey

¾ oz. sweet vermouth

¼ oz. maraschino liqueur

1 dash Angostura bitters

1 3-inch piece of caramelized rhubarb

Garnish: *brandied cherry, rhubarb, Aster*

Mint Julep

Classic

2 oz. bourbon
¾ oz. mint syrup

Pro tip: *Smack the mint for the garnish against your hand or the counter a couple of times to help release its aromatics.*

The Mint Julep has a long history, beginning in the Persian Empire with a medicinal rosewater drink called a gulab. Mint juleps made with rum or Cognac were popular in the American South as early as the 1700s, when they were often consumed in the morning. In the 1900s, bourbon took over as the most common and affordable spirit in the US, and the Kentucky Derby classic as we know it today was born.

Some Mint Julep recipes call for muddled mint and simple syrup, but I like making a mint syrup to keep the preparation simple and the drink free of small pieces of mint.

Combine the bourbon and the mint syrup in a rock glass or a julep cup. Add crushed ice, piling it slightly above the rim of the cup. Serve with a straw.

Mint Syrup: see pg. 145

Garnish: *mint, powdered sugar*

Tea Thyme

Mocktail

The flavors of ripe peaches and sweet black tea in this mocktail evoke the same spirit of summer in the South as a Mint Julep. Making a syrup by blending peaches with honey instead of simmering them on the stove helps to preserve their fresh, juicy flavor.

Combine ingredients in a julep cup or rocks glass and stir briefly. Fill with crushed ice, piling it above the rim. Serve with a straw.

Peach Syrup: Combine ingredients in a blender and blend until liquefied. Pour through a fine strainer.

2 oz. cooled black tea

¾ oz. lemon juice

½ oz. peach-thyme honey syrup

Peach Syrup:

1 peach, sliced (about 1 cup)

6 tbsp. honey

1 tbsp. lemon juice

¼ tsp. fresh thyme leaves

Garnish: *peach, thyme, micro princess flower*

New York Sour

Classic

2 oz. rye or
bourbon whiskey

1 oz. lemon juice

1 oz. simple syrup

1 oz. red wine

The layered New York Sour looks incredibly complicated to make, but in reality it couldn't be easier. It's my favorite cocktail to really wow my guests.

Combine whiskey, lemon juice, and simple syrup in a shaker with ice and shake until chilled. Strain into a rocks glass filled with ice. Slowly pour the red wine into drink over the top of a spoon to create the float.

New York Sour Popsicles

Original

Boozy popsicles are a fun summer treat. These are layered like a New York Sour for an impressive effect.

Combine the first layer ingredients in a mixing glass with a spout and stir. Divide among ten popsicle molds, filling them about ¾ full. Place in freezer for two hours, and then insert popsicle sticks if your molds use them. Let freeze overnight.

Combine the second layer ingredients in a mixing glass with a spout and stir. Add to the popsicles, filling to the top of the molds. Let freeze several hours or overnight. Run under warm water before gently removing popsicles from the molds.

First layer:

2 oz. rye

3 oz. lemon juice

3 oz. simple syrup

10 oz. water

Second layer:

6 oz. red wine

3 oz. water

½ oz. simple syrup

Garnish: *violas*

Old Fashioned

Classic

2½ oz rye or
bourbon whiskey

1 tsp. simple syrup

2 dashes
Angostura bitters

*The Old Fashioned is the original cocktail,
a simple mixture of spirit, sugar, water,
and bitters.*

Stir all ingredients with ice and strain into
a rocks glass over one large cube. Twist an
orange peel over the drink and either drop it in
or discard.

Garnish: *calendula*

Fall Fashioned

Original

Whiskey drinks naturally lend themselves to fall flavors like cinnamon, clove, allspice, apple, and cranberry. This twist on an Old Fashioned is perfect for a chilly day.

Stir all ingredients (including cinnamon stick) with ice and strain into a rocks glass over one large cube. Twist an orange peel over the drink and either drop it in or discard it.

2 oz. bourbon

1 oz. apple cider

1 tsp. maple syrup

1 cinnamon stick

3 dashes Angostura bitters

Garnish: *cinnamon stick, red hibiscus leaf*

Whiskey Smash

Classic

2 oz. rye or
bourbon whiskey

¾ lemon juice

¾ simple syrup

5 mint leaves

Pro tip: *Remember not to muddle your mint too forcefully; you want to gently release its oils, not tear it apart.*

A whiskey smash is a type of julep and used to be made very similarly to today's Mint Julep. Bartender Dale DeGroff resurrected the drink at the iconic Rainbow Room in New York with the addition of lemon juice, and it became an instant classic.

Muddle mint, lemon, and simple syrup, add whiskey, shake with ice, fine strain into a glass filled with crushed ice and heap more on top. Serve with a straw.

Garnish: *mint, thyme flowers*

Ginger Sage Smash
Original

The smash is a template with infinite possibilities. This version swaps out the mint for sage and makes the addition of fresh ginger. Try getting creative with your own flavor combinations.

2 oz. rye or bourbon whiskey

¾ oz. ginger syrup

¾ oz. lemon

4–5 sage leaves

Combine sage leaves and ginger syrup in the bottom of a shaking tin and gently muddle. Add whiskey and lemon juice and fill the shaker with ice. Shake until chilled and strain into a julep cup or rocks glass filled with crushed ice. Heap more crushed ice on top and serve with a straw.

Ginger Syrup:

½ cup sugar

½ cup water

1 inch of fresh ginger, peeled and sliced

Ginger Syrup: Bring to a simmer, stirring until sugar is dissolved. Add ginger and continue to heat, stirring, for one minute. Then turn off heat, cover, and let sit until cooled. Fine strain before using.

Garnish: *mint, borage*

Index

Acknowledgments

This book would not exist without the encouragement and creativity of the cocktail community on Instagram. Thank you for the endless inspiration, conversation, and support. In particular, I raise a glass to my Boston #Drinkstafam, for their friendship and everything they've taught me about spirits, cocktails, and the hospitality industry.

Thank you to Gourmet Sweet Botanicals for providing many of the fresh garnishes for the cocktails I shot before my garden bloomed.

Thank you to my parents for their pride and encouragement, and for employing their expert grandparenting skills when I needed them most.

And finally, thank you to my husband Tommy. You have always accepted my ups and downs, encouraged my myriad interests, and believed I could do something great. I love you.

About the Author

Katie Stryjewski is a writer, cocktail photographer, recipe developer, and Instagram influencer. She trained as an ornithologist and evolutionary biologist, receiving her PhD from Boston University and completing a postdoc at Harvard University before transitioning to her current career. She lives in Somerville, Massachusetts, with her husband and son. You can find her on Instagram as @garnish_girl and visit her blog, Garnish, at www.garnishblog.com.

Mango Publishing, established in 2014, publishes an eclectic list of books by diverse authors—both new and established voices—on topics ranging from business, personal growth, women's empowerment, LGBTQ studies, health, and spirituality to history, popular culture, time management, decluttering, lifestyle, mental wellness, aging, and sustainable living. We were recently named 2019 *and* 2020's #1 fastest growing independent publisher by *Publishers Weekly.* Our success is driven by our main goal, which is to publish high-quality books that will entertain readers as well as make a positive difference in their lives.

Our readers are our most important resource; we value your input, suggestions, and ideas. We'd love to hear from you— after all, we are publishing books for you!

Please stay in touch with us and follow us at:

Facebook: Mango Publishing
Twitter: @MangoPublishing
Instagram: @MangoPublishing
LinkedIn: Mango Publishing
Pinterest: Mango Publishing
Newsletter: mangopublishinggroup.com/newsletter

Join us on Mango's journey to reinvent publishing, one book at a time.